Excavations on Copa Hill, Cwmystwyth (1986-1999)

An Early Bronze Age copper mine within the uplands of Central Wales

Simon Timberlake

with contributions by

T. Mighall, S. Clark, A. Caseldine, N. Nayling, D.M. Goodburn,
B. Craddock, J. Ambers, A.E. Annel and R.A. Ixer

BAR British Series 348
2003

Published in 2016 by
BAR Publishing, Oxford

BAR British Series 348

Excavations on Copa Hill, Cwmystwyth (1986-1999)

ISBN 978 1 84171 486 8

© S Timberlake and the Publisher 2003

The author's moral rights under the 1988 UK Copyright,
Designs and Patents Act are hereby expressly asserted.

All rights reserved. No part of this work may be copied, reproduced, stored,
sold, distributed, scanned, saved in any form of digital format or transmitted
in any form digitally, without the written permission of the Publisher.

BAR Publishing is the trading name of British Archaeological Reports (Oxford) Ltd.
British Archaeological Reports was first incorporated in 1974 to publish the BAR
Series, International and British. In 1992 Hadrian Books Ltd became part of the BAR
group. This volume was originally published by Archaeopress in conjunction with
British Archaeological Reports (Oxford) Ltd / Hadrian Books Ltd, the Series principal
publisher, in 2003. This present volume is published by BAR Publishing, 2016.

Printed in England

BAR titles are available from:

            BAR Publishing
            122 Banbury Rd, Oxford, OX2 7BP, UK
EMAIL   info@barpublishing.com
PHONE +44 (0)1865 310431
FAX   +44 (0)1865 316916
            www.barpublishing.com

EXCAVATIONS ON COPA HILL, CWMYSTWYTH (1986-1999) AN EARLY BRONZE AGE COPPER MINE WITHIN THE UPLANDS OF CENTRAL WALES

By SIMON TIMBERLAKE
with contributions by T. Mighall, S. Clark, A. Caseldine, N. Nayling, D.M. Goodburn, B. Craddock, J.Ambers, A.E.Annel and R.A.Ixer

Figure 1: Mining on Copa Hill in the Bronze Age. Impression by B.Craddock 1991

ABSTRACT

The Comet Lode Opencast and the surrounding prehistoric-modern archaeological landscape on Copa Hill has been the subject of a major long term investigation by the Early Mines Research Group. This Early Bronze Age opencast is one of 12 now identified within Central Wales Orefield, but is almost certainly the best preserved, and probably the most fully excavated example of this class of upland primitive trench mines within the British Isles. More than 10 years of excavation here has revealed an intact 5-6 metre deep Early Bronze Age mining stratigraphy preserved under semi-waterlogged conditions within the base of the working, with an abundance of stone, antler, and wooden mining artefacts in situ., including some of the earliest recognized examples of mine drainage equipment (wooden launders) ever found. It appears that the earliest exploitation of the copper-lead ores may have begun sometime before 2000 BC following prospection activity within the Ystwyth Valley, with activity reaching its maximum almost one hundred and fifty years later (by which time the opencast had been mined to a depth of over 12 metres), thereafter continuing intermittently up to the point of its final abandonment around 1600 BC as a result of the exhaustion of near-surface deposits of chalcopyrite (copper) ore, as well as an increasing problem with flooding. This had affected some of the deepest parts of the mine which quite early on had been abandoned whilst most of the upper reaches continued to be worked. Mining by firesetting, using brushwood oak fuel and hafted stone cobble tools brought from the valley bottom and from the coast took place on a seasonal basis, usually in the spring/summer months. Such periodicity and lack of settlement remains may suggest that mining here was a secondary occupation of pastoralists, and as such the model for Early Bronze Age mining and prospection within the upland zone may well contrast with the slightly later and more long-lasting opencast/underground exploitation of the copper deposits upon the Great Orme's Head, Llandudno in North Wales. Of particular interest also is evidence which suggests that lead ores were also being systematically removed from the veins, and that in some cases these ores were being crushed and separated, but also apparently discarded. This therefore raises important questions about metallurgical experimentation, and/or the first use of lead, or leaded bronze alloying in Britain.

Important palaeo-environmental data has also been obtained from an examination of the sequence of infilling peats and silts which seal these early mining deposits, as well as from cores taken from the blanket peat above the mine. These have provided hitherto unavailable evidence as to the history of local woodland clearance, agriculture, and the record of prehistoric - modern mining and metallurgical activity within an area of the uplands sparse in extant archaeological remains.

CONTENTS
page

CHAPTER 1 **Introduction: History of research into Bronze Age Mining**.................................. 1
CHAPTER 2 **Site location: Geomorphology and geology**
        The site………………………………………………………………………………. 4
        Geology, mineral deposits and veins... 5
CHAPTER 3 **Mining at Cwmystwyth and discoveries of stone tools**
        The historical record of mining.. 7
        The record of modern mining within the vicinity of the Comet Lode Opencast and discoveries of stone tools.. 9
        Former archaeological work... 10
CHAPTER 4 **Landscape survey and archaeological investigations on Copa Hill (1986-1999)**
        Prehistoric mining... 11
        Prehistoric burial.. 14
        Medieval mining... 15
        Medieval agricultural and dwelling.. 16
        Early postmedieval mining... 18
        Victorian and modern mining... 20
        Victorian and modern agriculture.. 20
CHAPTER 5 **Excavations in the vicinity of the Comet Lode Opencast (1986)**
        Area D1... 23
        Area C1... 25
        Area A1... 26
CHAPTER 6 **Excavations within the prehistoric opencast mine (1989-1999)**
        1989 - Areas D2 and D3... 27
        1990-1993 - north side of opencast and mine gallery... 31
        1993-1999 - excavations at front of opencast.. 36
        Area D7 - mine infill etc... 36
        Area D8 - Mine Entrance A... 39
        Area D7 - Deep Fissure 1... 46
        Areas D9 - D11 - opencut trenches at front of opencast... 48
        Excavations of the Lateral Spoil Tip (1994-1999) - the earliest mining evidence............... 51
CHAPTER 7 **Investigations of the area surrounding the opencast (1993-1999)**
        South-east of the prehistoric tips.. 52
        Geophysical survey... 52
CHAPTER 8 **Radiocarbon chronology** (J.Ambers).. 55
CHAPTER 9 **The palaeo-environmental evidence : plant remains, beetles and wood**
        Plant macro-fossils (A.Caseldine).. 60
        Beetles (S.Clark)... 61
        Wood and charcoal (S.Johnson and N.Nayling)... 63
CHAPTER 10 **The palaeo-environmental evidence : palynology and peat geochemistry** (T.Mighall). 66
CHAPTER 11 **Wooden artefacts**
        Launders and their function.. 69
        Other wooden artefacts... 71
CHAPTER 12 **Wood technology** (D.Goodburn)... 81
CHAPTER 13 **Antler finds**... 84
CHAPTER 14 **Hammer-stones** (S.Timberlake and B.Craddock)... 86
CHAPTER 15 **Ore mineralogy** (R.A.Ixer).. 99
CHAPTER 16 **Ores mined in prehistory**
        Mineral veins, ore grade and the extraction of lead and copper........................... 100
        Notes on ore processing (A.E.Annel).. 102
CHAPTER 17 **A summary of the sequence of exploitation and abandonment history**.................. 103
CHAPTER 18 **Discussion**
        The scale of mining... 107
        The early exploitation of lead at Cwmystwyth.. 108
        Copper and Early Bronze Age metalwork in Wales and beyond............................ 109
        The use and availability of resources and the impact of mining upon the environment....... 110
        The social organization and significance of mining in the context of use and occupation of the upland zone... 112

CHAPTER 19	**Summary conclusions**..	115
Acknowledgments..		117
BIBLIOGRAPHY..		118
APPENDIX 1	**Early Mines Research Group Standard Record Sheet for Hammer-stones**................	124
	Hammer-stone Data Record Key	
	Krumbein's (1941) visual roundness chart for cobbles..	127

ILLUSTRATIONS:

TABLES
                                                                                                      page

TABLE 1 C14 dates: sample context, dates and calibrated age ranges for samples (J.Ambers)............ 55-58
TABLE 2 C14 dates: calibrated date ranges shown as probability distributions (J.Ambers)................. 59
TABLE 3 Alpha-numerical classification for hammer-stone function/modification (S.T.).................. 89

COLOUR PLATE

PLATE 1

(A) Mining landscape on Copa Hill - 5000 years of industrial activity
(B) Hushing channels between the Comet and Kingside Lode
(C) Recording > 4 metres depth of scree, silt, peat and spoil infilling the Bronze Age workings (1993)
(D) Excavations at the front entrance to the prehistoric mine (1995)
(E) A withy handle coming to light during excavations carried out within the Mine Entrance in 1994.......frontspiece

FIGURES (B&W)

FIGURE 1   Mining on Copa Hill in the Bronze Age (an impression by B.Craddock)....................... title page
FIGURE 2   Sites of prehistoric mines in Wales (map)... 2
FIGURE 3   Map of mineral veins, prehistoric mines and prospection areas in mid-Wales...................... 3
FIGURE 4   View of Cwm Ddu from Comet Lode Opencast... 4
FIGURE 5a   Copper staining on the walls of the Copper Level, Copa Hill.. 5
FIGURE 5b   Schematic (longitudinal) section of Comet Lode Opencast plus modern workings beneath... 6
FIGURE 5c   Modern ore-dressing mill (c.1900) at Cwmystwyth Mines (photographed in 1966)............... 6
FIGURE 6   W.W.Smyth's 1847 map of Cwmystwyth Mines... 8
FIGURE 7   Detail from O.T.Jones' 1922 map of Copa Hill... 9
FIGURE 8   Details from O.Davies' map and sections of trenches, Comet Lode Opencast (c.1937)........ 10
FIGURE 9   Comet Lode Opencast before excavation (1986)... 11
FIGURE 10   Location map (Cwmystwyth) and survey of Copa Hill (EMRG 1989)............................... 12
FIGURE 11   Prehistoric tips dissected by hush gullies below Comet Lode Opencast............................... 13
FIGURE 12   Early hushing activity on Copa Hill.. 13
FIGURE 13   Pant Morcell cairn.. 14
FIGURE 14   Course of ancient (Medieval?) leat along brow of Copa Hill... 15
FIGURE 15   Penguelan lead smelting site: excavations and geophysical survey results........................... 17
FIGURE 16   Platform house, Nant Stwc... 16
FIGURE 17   Plan of Nant Stwc early Medieval platform.. 18
FIGURE 18   Dry-stone walled ore-dressing shelters with anvil stones (Postmedieval), Kingside Lode..... 19
FIGURE 19   Probable 18th-century hushing dam on Copa Hill.. 19
FIGURE 20a Plan of turbary stacks and peat tracks in vicinity of Comet Lode Opencast......................... 21
FIGURE 20b Contour plan of Comet Lode Opencast showing excavation trenches and other features..... 22
FIGURE 21   View of trench D1 (across Central Tip), 1986... 23
FIGURE 22a Trench D1 (drawn section 1986).. 24
FIGURE 22b Photo-montage of section showing hammer-stones recovered.. 24
FIGURE 22c Fragments of decayed antler found in mine tip (trench D1a).. 24
FIGURE 23   Excavated channel in trench C1... 25
FIGURE 24a Wall tumble on foundations of turbary (peat-drying) stack with 'saddle-quern type' mortar. 26
FIGURE 24b 'Saddle-quern type' mortar stone/anvil - detail 26
FIGURE 25   Section across buried channel in trench C1 (drawing).. 26
FIGURE 26   Remains of stone sluice in base of early hushing dam (front of opencast: Area D2).............. 28
FIGURE 27a South section Area D2 showing turf-stack hushing dam in x-section.................................... 28
FIGURE 27b South-east sondage of D2 - section through organic infill... 28
FIGURE 28   Exposing prehistoric mine gallery against north face of opencast (Area D3)........................ 29

FIGURE 29a	Roof of mine gallery and imprint of stone tools (drawing B.Craddock 1989)	30
FIGURE 29b	Photographic detail of tool-marks within roof of mine gallery	30
FIGURE 29c	Plan plus drawn profiles across mine gallery	30
FIGURE 30	Recording inside of mine gallery in 1990	31
FIGURE 31	Backfilled shaft feature seen within infill layers (Area D3)	32
FIGURE 32	Metal pick holes within rock wall below level of mine gallery	32
FIGURE 33	Small cache of stone tools found *in situ*. within mine gallery (1990)	32
FIGURE 34	Mined-out hollow inside of gallery - shows tool marks and small spring (1990)	33
FIGURE 35	Fragments of oak mortice and tenon joint associated with lining of Postmedieval shaft	33
FIGURE 36	Sections recorded across Area D3, including upon floor of mine gallery	34
FIGURE 37	Plan of all excavation cuts and sections within opencast (1986-1999)	35
FIGURE 38	South-east side of opencast plus top of Rock Bench 1	36
FIGURE 39	North end of exposed alder launder *in situ*, 1993	37
FIGURE 40	Recording upper section of mine infill (D7 Deep Infill section)	37
FIGURE 41	Running section (drawing) of D7 Deep Infill Section	38
FIGURE 42	View of excavated Deep Fissure 1, Mine Entrance A and launder	39
FIGURE 43	Brushwood infill and support for base of launder in Entrance A	40
FIGURE 44	Detail of south section (length) of alder launder	40
FIGURE 45	Main x-section (drawing) through infill of Entrance A and Vein Fissure 3	41
FIGURE 46	Plan of floor of Entrance A showing position of finds and x-sections (excavation baulks)	42
FIGURE 47	Sections from north to south across Entrance A	43
FIGURE 48	Plan of wood and withy finds within layer 055, sector A/C	44
FIGURE 49	Plan of timber debris deposited in Deep Fissure 1	44
FIGURE 50	Excavation in progress at south end of Entrance A (1995)	45
FIGURE 51	Fragments of coarsely woven hazel basket in process of being excavated (Entrance A)	45
FIGURE 52	Handle of antler hammer/pick projecting from base of section (A/D)	46
FIGURE 53	Primitive pick or chisel holes within galena vein (Fissure 1)	46
FIGURE 54	Broken section of alder launder (024) discarded into Fissure 1	47
FIGURE 55	Recording within base of Fissure 1 excavation (1999)	48
FIGURE 56	Area D9 - Vein Fissure 3 fully excavated	48
FIGURE 57	Oak beam stemple (103) lying *in situ*. within Vein Fissure 3	49
FIGURE 58	South section of D9 (drawn)	50
FIGURE 59a+b	Sections (drawn) of trenches E5 and E7 in Lateral Tip (1999)	50
FIGURE 60	Placed stone slab (and excavated pit), Trench C6	53
FIGURE 61	Plan and south-west section (C6)	54
FIGURE 62	Fragment of bracken frond lying within Early Bronze Age mining layer	60
FIGURE 63	Chart of beetle habitat groups represented within peat infill of mine (S.Clark)	62
FIGURE 64	Tree species/ genus percentages for total wood assemblage (N.Nayling)	63
FIGURE 65	Hazel twigs lying on surface of mining floor (Entrance A)	64
FIGURE 66a	Scatter diagram of ring counts and average width of oak samples from 081(N.Nayling)	65
FIGURE 66b	Scatter diagram of ring counts and average width of oak samples from 053(N.Nayling)	65
FIGURE 67a	Pollen diagram and geochemical profile (copper) from core site CH2 (T.Mighall)	67
FIGURE 67b	Geochemical profiles for lead and zinc from monoliths CH1-CH4 (T.Mighall)	68
FIGURE 68	Section of alder launder *in situ*. (Entrance A)	70
FIGURE 69	Detail of alder launder (south end)	70
FIGURE 70	Excavated launder lying supported within re-cut ditch	70
FIGURE 71	Two halves of broken withy handle for hammer-stone	72
FIGURE 72	Twisted withy tie associated with basket (as excavated)	73
FIGURE 73	Sketch of basket from Esgairmwyn (drawn by Lewis Morris 1752)	73
FIGURE 74	Traditional faggot bundle and withy tie (after Tabor 1994)	73
FIGURE 75a	Wooden artefacts : alder launder (047) - drawing compiled from excavated sections	74
FIGURE 75b	Damaged alder launder section (024) plus large oak launder (083) - drawings	75
FIGURE 76	Oak stemples and mine timbers (drawings)	76
FIGURE 77	'Handled' sticks and wooden stakes (drawings)	77
FIGURE 78	Oak chips (drawings)	78
FIGURE 79	Twisted withy handle, ties and rope (drawings)	79
FIGURE 80	Fragment of basket (drawing)	80
FIGURE 81	Reconstruction of shallow basket type (B.Craddock)	80
FIGURE 82a	Suggested flat metal axe type used in woodworking	82
FIGURE 82b	Detail of axe stop-marks on alder launder (A.Gwilt)	82

FIGURE 83	Diagram(s) of parent alder tree and suggested stages in construction of launder................	83
FIGURE 84	Antler artefacts from Entrance A - pick/hammer and tine (drawings)...................................	85
FIGURE 85	Hammer-stone in mine spoil associated with charcoal and wood (photo)............................	88
FIGURE 86	Length/width plot for Copa Hill hammer-stones..	89
FIGURE 87	Weight (kg) frequencies for Copa Hill hammer-stones...	89
FIGURE 88	Chart of cobble shape frequencies for Copa Hill hammer-stones..	90
FIGURE 89	Cobble smoothness frequencies for selected cobbles...	90
FIGURE 90	Suggested tool function(s) within analysed assemblages of used cobbles (pie chart)............	90
FIGURE 91	Some examples of experimentally hafted hammers (1989)..	92
FIGURE 92	Stone tools from Copa Hill (drawings): end-hammers (little modified) A-F.......................	94
FIGURE 93	Stone tools (drawings): end-hammers, picks etc. (notched and semi-grooved) A-F............	95
FIGURE 94	Stone tools (drawings): flake tools (chisels etc.), waste flakes and spalls A-I.....................	96
FIGURE 95	Hand crushing/ pecking tools A-G...	97
FIGURE 96	Anvil stones, 'saddle quern type' mortar, and miscellaneous (stone lid) A-E......................	98
FIGURE 97	Conjectured plan of mineral veins and faults within front of opencast................................	101
FIGURE 98	Polished ore sections of chalcopyrite from both crushed ore and mine spoil.......................	102
FIGURE 99	Small galena vein (VF2) with discarded lead ore and stone chisel.....................................	102
FIGURE 100	Possible sequence of mining from prospection to abandonment. Reconstructions..............	105
FIGURE 101	Conjectured sequence of tipping outside of mine..	106
FIGURE 102	Map of Early Bronze Age field monuments within the Plynlimon area..............................	114
FIGURE 103	Schematic cross-section through centre-front of opencast showing mining and abandonment horizons..	116

PLATE 1 (A) + (B) Views of mining landscape on Copa Hill
Opencast (in sunlight)[TOP (A)]; hillside dissected by hushing channels, Comet Lode to right [BOTTOM (B)]

PLATE 1 (C - E)   Excavation in progress
Recording deep infill sections within opencast in 1993 (TOP LEFT (C)); excavations underway within entrance to mine in 1995 (D); the excavation of a withy handle (for a hammer-stone) in 1994 (E)

CHAPTER 1

INTRODUCTION

HISTORY OF RESEARCH INTO BRONZE AGE METAL MINING

Ideas on the antiquity of metal mining in Britain and the recognition of prehistoric mines are not solely the prerogative of modern archaeologists, archaeo-metallurgists and geologists, but more often than not owe their origin to the visits of mineral surveyors and antiquarians who witnessed the re-opening of old abandoned workings at the beginning of the modern industrial period. Thus in 1744 the Crown Mineral Agent Lewis Morris describes his workmen finding the remains of stone mining tools in opening up the old Twll y Mwyn mine near Penrhyncoch in mid-Wales, suggesting that it was "..wrought in the beginning of times, and before the use of iron was found out" (Bick & W.Davies 1994, 37), whilst a little later, during the great working of the Parys Mountain Copper Mine on Anglesey in 1796, Sir Christopher Sykes in his journal also described similar remains as being of a date "..before Iron was used in this Kingdom" (Sykes 1796). Elsewhere, a much more thorough analysis of ancient pit workings and the caches of stone tools found associated with a cupriferous Triassic sandstone uncovered as a result of modern mining at Brynlow on Alderley Edge in Cheshire in 1874 was provided by Prof. W. Boyd Dawkins in 1874. His conclusion was that these represented a (mining) phase "..which may point back to the bronze age, when the necessary copper was eagerly being sought throughout the whole of Europe" (Boyd Dawkins 1875, 79).

During the period of the Victorian mining boom, reports on the discovery of ancient mines, as well as 'Roman' shafts and levels became commonplace both within the pages of the Mining Journal (Timberlake 1991,179)and in specialist geological reports (Smyth 1848), resulting in an accumulated body of evidence which prompted the first systematic archaeological survey of ancient mining as late as the 1930's. The latter was undertaken by Oliver Davies on behalf of a British Association for the Advancement of Science (Section H) following the establishment of a committee in 1935 to investigate the evidence for early mining in Wales (Davies 1936; 1937;1938). Cwmystwyth was one of the first mines that Davies visited.

The presence of stone mining hammers was confirmed at many of these sites (including Cwmystwyth), yet in the absence of any other dateable artefacts or any independent means of dating, Davies ended up comparing these mines to those of known Roman date in which such tools had turned up in disturbed context (Davies 1937; 1939; 1947; 1948). He concluded that some or all of this work may have been carried out by the 'Celtic' inhabitants under some sort of overall Roman control - a rather circular argument based on the premise that it was Roman interests which had opened up these mining areas in the first place. Such views on the origins of mining persisted until quite recently (Shepherd, 1980), whilst even in archaeological circles the lack of any clear-cut evidence for the prehistoric exploitation of metals in Britain led to many assumptions being made regarding the use of Irish or continental sources of copper during the British Bronze Age (Coghlan & Case 1957). Some indeed had argued that suitable ore deposits were lacking (Muhly 1987), whilst others have maintained that sufficient ore for Bronze Age needs could have been scavenged from superficial deposits on the surface (Briggs 1988; 1991). However, strong suspicions of Bronze Age mining activity on the mainland continued to proliferate (Richardson 1974; Jones 1979), fuelled no doubt by the first C14 date from the primitive 'stone hammer' mines on Mount Gabriel, Co.Cork in Ireland (Jackson 1968, 92-114). Yet it was not until 1986 that the first proof of this appeared, following underground reconnaissance on the Great Ormes Head, Llandudno in North Wales (James 1988).

Similar investigations were being undertaken concurrently by the author and others (Pickin 1988; Craddock 1986; Gale, 1986), and within the space of a few years number of other probable sites had been identified, and several preliminary excavations, such as those carried out at Cwmystwyth (Timberlake 1987; Timberlake & Switsur 1988), Parys Mountain, and Nantyreira near Plynlimon (Timberlake 1988) had all returned Bronze Age radiocarbon dates from sections cut through undisturbed ancient mine spoil. Around this time the Early Mines Research Group was formed to investigate mining in Wales and beyond (but still with a specific emphasis on Central Wales which contains the largest concentration of identified Early Bronze Age metal mines in Britain). On the Great Ormes Head excavations carried out since 1988 by the Gwynedd Archaeological Trust, the Great Orme Mines Ltd., and the Great Orme Exploration Society (Dutton & Fasham 1994; Lewis 1994; 1996; David 1996; 1997;1998) have uncovered what would appear to be one of the largest complexes of surface and underground Bronze Age (EBA - LBA) mining in Europe. Other more recent investigations such as on Parys Mountain (Jenkins 1995) also seem to suggest that mining carried out here during the Early Bronze Age (which includes the underground exploitation of copper sulphide ores) was probably much more extensive than had once been imagined. Meanwhile, six Early Bronze Age mines working copper or copper-lead veins had now been identified and dated (Timberlake 1992; 1995; 1996; Timberlake & Mason 1997) in Central Wales. Here the Bronze Age miners appear to have exploited only small sections of what sometimes must have been richly mineralized veins at surface, in some cases ignoring rich ore, and 'flitting from one shallow opencast or pit to the next' (Peake 1937; Timberlake

2001, 182). Indeed, apart from Cwmystwyth, most of these operations appear to have been little more than very small-scale workings or prospections that we find located on the sides of major valleys into the interior (Tyn y Fron, Nantyrarian) or else seaward-facing around the Dovey Estuary (Llancynfelin, Ogof Wyddon). Research is currently on-going to further identify these prospection areas and to relate this to what we know of contemporary (agricultural) exploitation and occupation within the uplands (Timberlake 2002).

Currently the search for the sources of Bronze Age metal has shifted to investigating possible sites within North, Central and SW England. To date, no Bronze Age copper mines have been identified within metalliferous province of the SW Peninsula (Craddock & Craddock 1996;1997; Sharpe 1997; Budd & Gale 1994), although fieldwork and excavation carried out at sites within Southern Pennines (Alderley Edge, Cheshire and Ecton in N.Staffordshire) has proved rather more successful. At Alderley Edge a Bronze Age oak mining shovel recovered in 1878 was recently C14 dated (Garner et al. 1994), prompting excavations at Engine Vein which have since uncovered an undisturbed Early Bronze Age pit-working (Pain 2001; Timberlake (2003 *forthcoming*)), whilst on the top of Ecton Hill the find of hammerstone tools and part of a 3700 year old antler mining pick underground have also helped to identify several areas of prehistoric mining (Barnatt & Thomas 1998; Barnatt 1999; Pickin 1999). The plurality of sources of British copper used during the earlier part of the Bronze Age appears now to have been confirmed by analyses carried out on the lead isotopic ratios and on metal impurities present within both bronze artefacts and a wide range of potential ores (Rohl & Needham 1998; Northover 1980; 1983). Overall, there seems little doubt that many other Bronze Age and later prehistoric metal mines still remain to be discovered within England, Scotland and Wales.

With some justification the trench mine on Copa Hill might well be considered to be the most intact surviving example of a Bronze Age metal mine in Britain, in many respects reminiscent of the near completely infilled shafts and galleries of the Neolithic-Early Bronze Age flint mines at Blackpatch, Harrow Hill, Hambledon Hill and Grimes Graves (Holgate 1991). However, much more direct comparison can now be made with the more thoroughly investigated primitive copper mines on Mount Gabriel (O'Brien 1987; 1990; 1994), which are likewise peat filled and waterlogged, and which have yielded a very similar assemblage of cut and burnt wood fuel, wooden artefacts, and un-grooved stone mining tools. A still closer parallel with this type of vein fissure trench mine may be the ancient working discovered during peat-cutting operations at Derrycarhoon, also in Co. Cork, Eire during the early 19th century (Jackson 1968; O'Brien 1989). The latter trench mine (18 m long by 18 m deep and approximately 1.5 m wide) was also found buried beneath 4 metres of peat. Unfortunately this site was almost completely destroyed during subsequent mining operations, thus the opportunity to excavate and accurately date such a mine was lost. However, the surviving remains of Chalcolithic - Late Bronze Age metal mines from almost every other part of Europe show certain basic similarities in style and method of working and tool use, and accounts of these compare well (SEE: O'Brien 1995 (Ross Is. Eire); Lechelon 1974 (Bouco-Payrol, France); Rothenberg & Blanco-Freijeiro 1981(Huelva, Spain); Pittioni 1951; Gale & Ottaway 1990 (Mitterberg, Austria); Mussche et al. 1990 (Thorikos, Greece); Bogosavljevic 1988 (Mali Sturac, Yugoslavia)). Nevertheless, it is also becomes clear from the study of many other examples (some way beyond this geographical sphere of influence or chronological framework, for example those described by Herbert 1984 (throughout sub-Saharan Africa); Griffin 1961 (Keewanaw Peninsula, Lake Superior, N.America); and Bird 1979 (Chuquicamata, N.Chile)), that such apparent similarity in styles of exploitation may owe much more to the identical nature of the utilitarian mining response appropriate to the range of geological and material resources conditions faced by pre-iron using societies, than to the diffusion of new ideas or the long distance movements of peoples skilled in mining and prospecting.

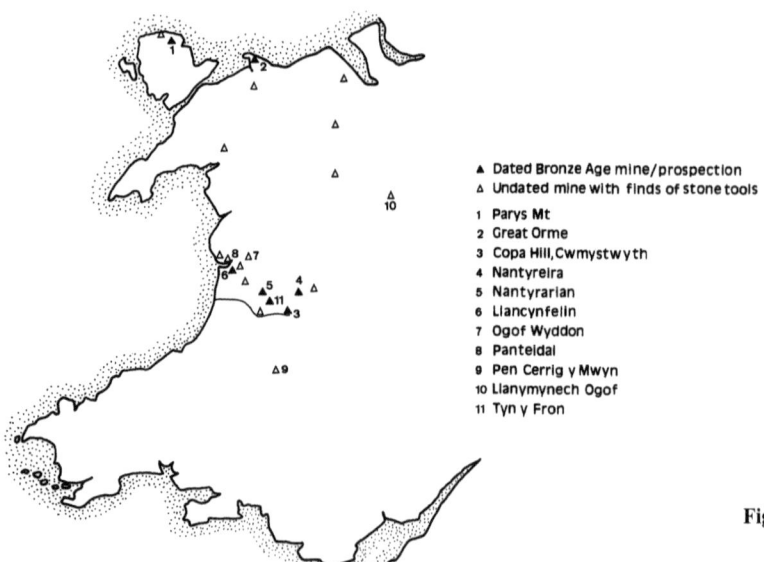

▲ Dated Bronze Age mine/prospection
△ Undated mine with finds of stone tools

1 Parys Mt
2 Great Orme
3 Copa Hill, Cwmystwyth
4 Nantyreira
5 Nantyrarian
6 Llancynfelin
7 Ogof Wyddon
8 Panteidal
9 Pen Cerrig y Mwyn
10 Llanymynech Ogof
11 Tyn y Fron

Figure 2: Sites of prehistoric metal mines in Wales

Figure 3: Map of prehistoric mines, mineral veins and prospection areas in mid-Wales

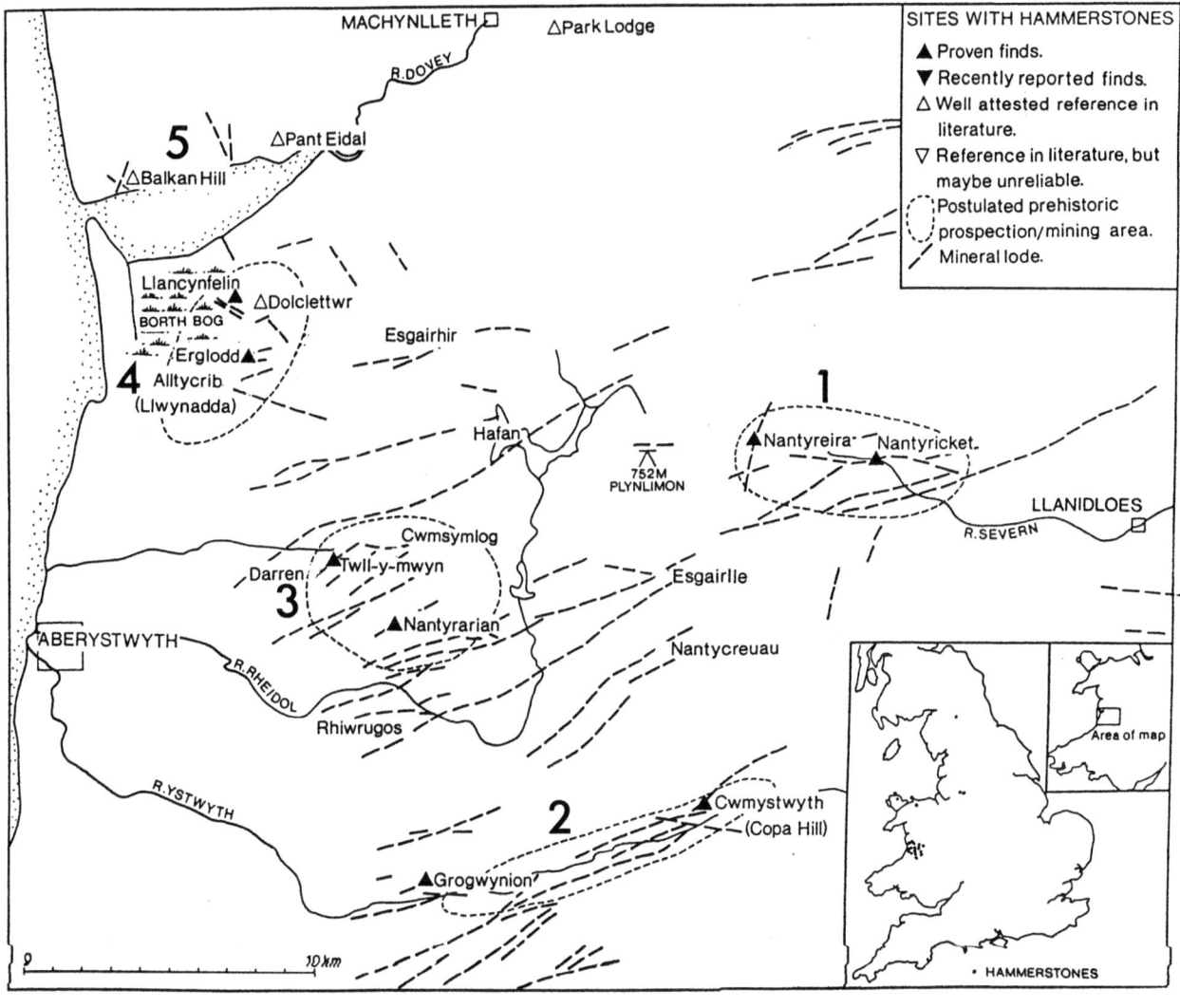

CHAPTER 2

SITE LOCATION: GEOMORPHOLOGY AND GEOLOGY

THE SITE

The Comet Lode Opencast [SN 81167520] is to be found at an elevation of 426 m OD, located at the top of an area of mine workings which covers the SW slopes of Copa Hill at the far eastern end of the Cwmystwyth Mines, on the north side of the Upper Ystwyth Valley. The site overlooks the small side valley of the Nant yr onnen to the west and the floor of the glaciated U-shaped Ystwyth some 180 metres below to the south, whilst to the north and east the land rises a further 70 metres to the edge of a dissected peat covered plateau. The opencast cuts a prominent rock outcrop on the brow of Copa Hill, below which the slopes are covered with a thick mantle of moraine drift, scree and solifluction deposits. A number of small streams and gullies descend to the valley floor of the Ystwyth, but on the gentler SW slopes of Copa Hill most of the visible gullies are artificially generated hushing channels excavated for the purpose of prospection and mining.

Near its confluence with the Ystwyth, a lobe of glacial moraine and alluvial deposits with large erratic boulders in its base fills the floor of the small Nant yr onnen valley. The latter forms a prominent terrace above the road, down through which the present stream has cut. Just west of this, where this stream meets the flood plain of the main river, a shallow cliff formed by the erosive power of the migrating channels of the Ystwyth has revealed several metres of alluvial deposit near its tip. The upper part of this consists of redeposited mine spoil, including fragments of ore, slag, and clinker. Almost opposite on the south side of the Ystwyth Valley lies the important geomorphological site of Cwm Du, an incipient cirque containing a series of nivation terraces or gelifluction fans formed as a result of snow-field melt during the late Devensian (Watson 1968; James 2001).

Valley-side moraine continues to fringe the Ystwyth upstream for another 5 km to its source, close to the county boundary at Maen Hir on the west side of the Cambrian Mountains, also the watershed with the Elan. The Ystwyth River originally had its mouth at Tan y bwlch, Aberystwyth, some 31 km to the west. The small village of Cwmystwyth lies about 1.5 km distant from the mines Raised peat mire (blanket bog) has developed on the plateau above the 470 m contour, some of this to a depth of 2-3 metres, and much of it exposed within the base of eroding peat hags. The vegetation of the drier moorland consists of heather, bilberry and purple moor grass, whilst cotton grass, sphagnum and sundew are commonly to be found within the wetter erosion gullies and bogs (Clark 1996).

Today, apart from a few rowan growing upon the steeper slopes, the valley remains treeless above the 250 m contour, consisting largely of grass covered hillside scree slope, and open and overgrown mine workings, which provide rough grazing land for sheep, and a haunt for buzzards, kites, crows and peregrines. To the west of the Nant yr onnen stream, the north side of the Ystwyth Valley projects to form the prominent spur of Pant Morcell, whilst further west and below lie the main 18th-19th century mine workings of Pugh's and Kingside Mines. On the crest of the rocky escarpment above this, in between the Nant Gwaith and Nant Watcyn streams, can be seen the scar of the Graig Fawr, the scene of considerable opencast and underground mining from Medieval times onwards. This whole complex of workings, over a linear distance of about 1.5 km is now protected as a Scheduled Ancient Monument, whilst parts of it (including Copa Hill) lie within a triple SSSI (Site of Special Scientific Interest).

Figure 4: View of Cwm Ddu from Comet Lode Opencast (S.Timberlake 1999)

GEOLOGY, MINERAL DEPOSIT AND VEINS

The outcropping mudstone, sandstone and gritstone rocks that form part of the Cwmystwyth Grits Group of Upper Llandovery, Silurian age (BGS Llanilar map sheet) are stratigraphically the highest beds within the Central Wales Orefield which have been intersected by significant vein mineralisation. At Cwmystwyth these veins form a major system, over 1.5 km in length, of mainly ENE-trending fractures (faults) which cut these synclinally folded strata, and within which the metal suphides of lead (galena), zinc (sphalerite), and to a lesser extent iron (pyrite) and copper (chalcopyrite) have been deposited. Hydrothermal mineralisation took place here in several repeated phases following fracturing which began in Devonian times and continued through into the Carboniferous (400-300 million years ago). The geology and mineralisation of this mine have been described in more detail by Jones (1922), Mason (1994; 1996;1997), Bevins and Mason (2001), Ixer & Budd (1999).

More than 30 different veins have been worked at various times, although three principal ones can be identified - the Comet (Belshazzar) and Kingside Lodes (both of which dip 50-65' to the south), and the less important northerly dipping Mitchell's Lode. Both the Comet and Kingside Lodes meet and join up along the Graig Fawr to form a particularly rich (lead/zinc) ore shoot, but these again diverge on Copa Hill, where the copper content of the Comet Lode increases eastwards, particularly in its upper part (Jones 1922). Within the vicinity of the Comet Lode Opencast this lode bunches and splits into a number of different ore shoots and forms a stockwork of lead and copper-bearing veins (Timberlake 1996; Mighall et al. 2000). The earliest (A1) type of mineralisation here can be seen within the Comet Lode (Mason 1994;1997), and this is represented by a series of hydraulic fractured mineral-cemented vein breccias (Phillips 1972). The first phase is cemented by quartz and sphalerite (A1-b), followed by quartz, galena, and chalcopyrite (A1-c/d), and then finally, quartz and abundant carbonate (A1-e), the latter consisting of ferroan dolomite or ankerite, a mineral which typifies this early mineralisation, and which has often been referred to elsewhere in mid-Wales as a sure harbinger of copper ore (Bick 1976,48). The later (A2) mineralisation has affected the Comet Lode to a somewhat lesser extent following re-activation of the fault(s), yet this type is more evident elsewhere at Cwmystwyth, where it has been responsible for much of the lead mined (Bevins & Mason 1997). The latter sequence consists of brecciated and crustiform-banded veins of sphalerite, galena, quartz, then finally late-stage pyrite and marcasite (iron sulphides) and calcite, but with copper invariably absent.

Secondary or supergene minerals are generally rare from Cwmystwyth, although lead minerals are commonly found amongst weathered tip material at the top of the Kingside Lode on Copa Hill (cerussite and pyromorphite), whilst in the Comet Lode, post-mining basic copper sulphates have been recorded on the walls of some of the levels (Bevins 1994). More recently a suite of secondary minerals including goethite, covellite, malachite, cerussite, anglesite, melanterite, sulphur, linarite, brochantite and langite have been identified as microscopic crystals within intensely altered crusts surrounding a galena-chalcopyrite-sphalerite vein exposed in the side of the prehistoric opencast (*pers. com.* T. Cotterell, National Museum of Wales). However, most if not all of this alteration is probably recent, there being no good evidence for the former presence of appreciable amounts of secondary copper minerals within the upper reaches of the Comet Lode. A more likely scenario is that the earliest exploitation took place within the sulphide zone, most probably for chalcopyrite, but perhaps also for galena (Mighall et al. 2000).

Figure 5a: Copper-staining (malachite and brochantite etc.) on the walls of the Copper Level located some 27 metres directly underneath Comet Lode Opencast (photo by R.Protheroe Jones 1988)

Figure 5b: Schematic section of Comet Lode Opencast (prehistoric workings) with modern workings beneath (at or above the 400 m contour). Drawing ST and BC 1995.

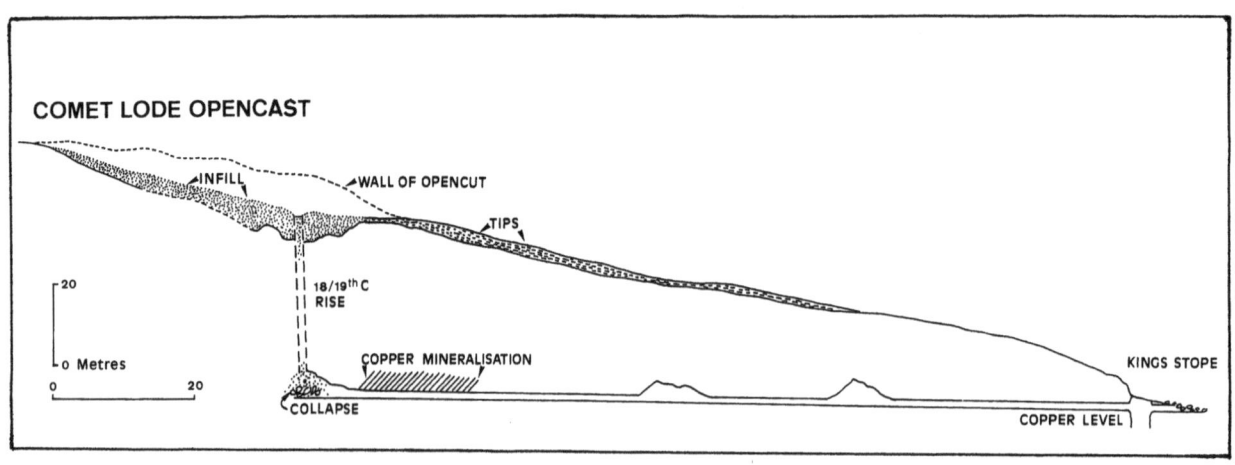

Figure 5c: The dressing mill on the valley floor at Cwmystwyth designed by Messrs. Buss and Davies and built for Henry Gamman in 1900. It survived until the early 1990's when it was dismantled as part of an ill-designed policy of landscape reclamation. Photo by Douglas Hague taken in July 1964.

CHAPTER 3

MINING AT CWMYSTWYTH AND DISCOVERIES OF STONE TOOLS

THE HISTORICAL RECORD: MINING AT CWMYSTWYTH AND IN THE UPPER YSTWYTH VALLEY

There is documentary evidence to suggest that both the mine and the lands thereabout belonged to the grange of Strata Florida Abbey (some 10 km to the south-west of Cwmystwyth) from at least the middle of the 12th century AD until the time of the dissolution in 1536, and circumstantial evidence too that the monks had mining interests both here and elsewhere in the area (Hughes, 1981). In 1887 evidence was found for monastic lead smelting and refining close to the abbey itself (Williams, 1889), whilst recent archaeological evidence from Copa Hill (this paper), suggests that lead was also being mined and smelted at Cwmystwyth during the 12th and 13th centuries AD. However, the date of the earliest surviving mining lease from Abbot Richard Talley of Strata Florida to Rhys and David ab Ieuan ab Hywel is probably no later than 1535 (Hughes 1981, 7). The first eye-witness account of mining here dates from around about the same time, and is that provided by John Leland in his approach along the old monastic road from Rhayader, during the course of his journey through Wales: "About the middle of this Wstwith valley that I ride in, being as I guess three miles in length, I saw on the right hand of the hill side Cloth Moyne (Clodd Mwyn = mine of lead), where hath been great digging for Leade, the smelting whereof hath destroid the woodes that sometimes grew plentifully thereabout" (Leland 1536-1539 IN Meyrick 1808). The description suggests that the working he witnessed was that around the area of opencasts on the Graig Fawr, a site referred to from the 17th century onwards as 'Craggie Moyne', a mine which it was later noted "..formerly belonged to the monastery of Strata Florida.."(Lewis Morris 1751). Yet it seems likely that the monks or their tenant miners were also working Copa Hill at this time , since following the suppression of Strata Florida, Henry VIII was of the opinion that all the land to the west of the Nant yr onnen was worthless, whereas he claimed Copa Hill for the Crown (Hughes 1981, 8).

Leases were issued in 1588 to Charles Evans and John Hopwood under the aegis of the newly formed Society of Mines Royal, and this earliest Elizabethan exploitation probably took place on or around the Graig Fawr. Thereafter a succession of leases including to Sir Hugh Myddleton and Sir Frances Bacon (1617), Thomas Bushell (1641), Edmund Goodyeare (1649-1660), and Morgan Herbert of the Hafod who worked a rich section of the Kingside Lode ('Herbert's Stope') on Copa Hill, suggest semi-continuous mining activity throughout the 17th century. This culminated in a lease of the mine to the newly formed Company of Mine Adventurers, by the then landowner William Powell of Nanteos, following the repeal of the Mines Royal Act in 1693.

Work here by the 'Mine Adventurers' under the management of William Waller appears to have involved considerable prospection work, but only a small amount of production, all of which was lead ore, and probably most of it from the old Kingside Lode workings on Copa Hill (Hughes, 1981). This period of activity is well documented, thus it is probably pertinent to mention that Waller only mentions 'Silver Hill' (possibly within the area of the Penguelan Mine on Copa Hill) but not 'Copa or 'Copper Hill', suggesting that the copper ores at the top of the Comet Lode at that time had not yet been re-exploited (Waller 1699; Hughes 1981). Thus this work probably took place between 1725 and 1740, at a time when mining generally was at a low ebb, and when small groups of miners were working on Copa Hill under a pretended lease from Thomas Powell (Morris 1751, 563). However, in 1749 at the time of John William's visit, the old opencasts on the Graig Fawr were then clearly being re-worked, and he refers to miners suspended in ropes, blasting down the walls with gunpowder(Williams 1780, 358). In 1751, Morris refers to 70 or 80 miners being constantly employed here, and the production of 200 tons of 'potters ore' per year (Morris 1751, 562).

A general economic revival and rising lead prices in 1758-1761 coincided with the interests of Chauncey Townshend who engaged the services of a Derbyshire man Thomas Bonsall to manage his interests here. Bonsall himself took the lease in the mid-1780's, for the first time exploiting the abundant deposits of zinc blende, no doubt stimulated by the new demand for this mineral following the perfection of the patent Champion process for zinc. Bonsall relinquished Copa Hill in 1791, and thereafter brought the mine into its most successful period, raising about 46 tons per month over the next 10 years, with a reputed income of £2-3000 per annum from Pugh's and Kingside Mines alone (Hughes 1981). Many of the most impressive remains of hushing at Cwmystwyth probably date from this period.

At the beginning of the 19th century work by small bands of miners centred on parts of Copa Hill, although short-lived success followed for a partnership of the Alderson Brothers from Swaledale (1822-1834), whilst rising metal prices also benefited Lewis Pugh of Aberystwyth who held the lease from 1835 to 1844. Pugh retained the services of the Yorkshire and Derbyshire miners and raised many thousands of tons of lead ore, chiefly from the western workings (Pugh's Mine), effectively commencing the first deep workings from the valley floor.

In 1848 John Taylor & Sons took over the controlling share of the mine and first introduced the modern Cornish style mining practice so typical of mid-late Victorian mines in Cardiganshire. A system of leats were cut, supplied by the headwaters of the River Ystwyth some 4 km upstream, to power waterwheels for drainage (pumping), the drawing of ore, and for crushing, whilst mill sites were established at both the Cwmystwyth (Pughs) and Kingside workings. Compressed air rock drills were introduced underground in the 1870's. Most of the large waste tips visible above the road date from this period of working, as do the inclines and tramways visible below the Graig Fawr and on Copa Hill.

Mining re-started in 1900 with a lease to Henry Gamman and the Cwmystwyth Mining Company, an event perhaps best remembered for the construction of the timber framed 'state of the art' ore dressing and concentrating mill, the remains of which survived on site until its final demolition in 1992. Pelton turbines fed by 740 feet head of water from a reservoir on the plateau above generated electricity and powered an air compressor plant, a tramway incline, and much of the mill machinery. The mine and plant were sold to a Belgian company Brunner Mond & Co. in 1909, and thereafter most of the capital was spent on modernisation, prospection, and development work, and little further production was seen, the very last return of ore for the mine being made in 1912. All activity at the mine ceased in 1923.

Interestingly, the final prospection work took place at the eastern end of the Comet Lode on Copa Hill, not a stones throw from the site of the earliest mining here some 3700 years before.

Figure 6: W.W. Smyth's map of Cwmystwyth Mines c.1847 showing mineral lodes and workings (from Hughes 1981).

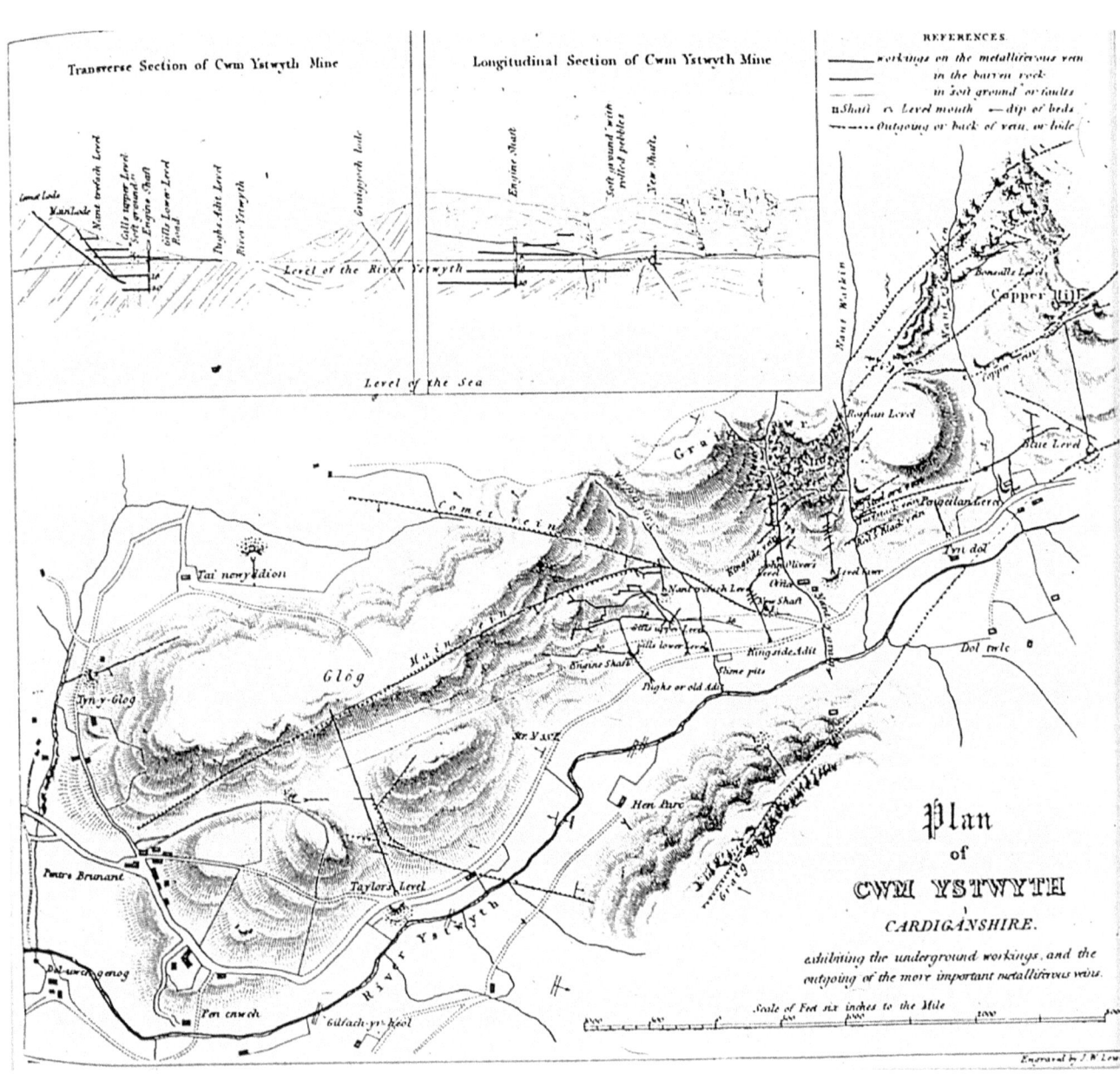

THE RECORD OF MODERN MINING WITHIN THE VICINITY OF THE COMET LODE OPENCAST AND THE DISCOVERY OF STONE TOOLS

Whilst there are few clear-cut references to work within this area of the mine, the change in the terms of the lease between William and Thomas Powell in 1723 and Ithelstain and Thomas Owen to include "all mines of lead and copper etc. in Cwmystwyth.." (Cwrt Mawr deeds NLW) may suggest that copper had been discovered and worked on the Comet Lode shortly before this. The likely date would have been between 1722-1723, since a lease granting Henry Parry to dig for ores on the lands of Nant yr Onnen (the former name of the smallholding on Copa Hill at the foot of the Comet Lode) had been granted the year before (Cwrt Mawr 208 NLW mss). Lewis Morris provides us with a rather more specific reference to "great quantities of copper ore formerly raised (on Copper Hill).....under a pretended lease from Mr.Powell" (Morris 1751, 563). The earliest reference to this place as 'Copper Hill', following the discovery of ore, would seem to suggest that the hill first got its name (or else was re-named as such) in the eighteenth century, and that the currently used name Bryn Copa [welsh = hill brow] is of much more recent origin . The latter, in fact, is such a commonly used name in Wales that subsequent cartographers may well have thought the original English wording to be a mistake (Davies 1947, 57). Whatever the origin, Lewis Morris omits to mention the existence of ancient workings at the head of the Comet Lode, even though he was well aware of the considerable antiquity of many of the Cardiganshire mines, a number of which he referred to as being "wrought by the Ancient Britons" (Morris 1756), some using stone tools (Morris 1744). It seems possible therefore that these particular workings were not uncovered before the end of the 18th century.

The earliest discovery within, or around the vicinity of the Comet Lode Opencast took place in 1813, when "..stones different from any in the neighbourhood were found in the old levels; round and of tough quality, and supposed by the present race of miners to have been used for hand mallets, before the use of pick-axes" (Davies 1815, 516). Copa Hill (Copper Hill Mine) at this time was being held by Joseph Jones of Blaen y Cwm, but in the same year he relinquished this lease to John Marsden who had already taken up Jones' former lease for the western half of the mine in 1811 (Hughes 1981). Indeed, Marsden himself may well have been responsible for this discovery (of the stone implements), since his miners had been carrying out development work and investigating old workings elsewhere at Cwmystwyth in 1813, within one of which they had found the remains of the old pumps (Davies 1815). A little later, the geologist Warrington Smyth refers to the discovery of stone tools inside old open copper workings on Copa Hill in a footnote to his survey of the mines published in 1848 (Smyth 1848, 664).The tools he describes here were "..hard ellipsoidal stones..most considerably worn at one end, and some marked by a rough groove round the middle, by which they seem to have been fastened to a handle". Francis (1874, 145-146) talks in more general terms about the occurrence of these hammer-stones both at Cwm Daren and on Copa Hill, noting their association with shallow opencasts dug along the veins where copper ore comes to surface, the tools themselves being waterworn pebbles some 6-7 inches long, 3-4 inches wide, and about 2 inches thick obtained from the sea shore, with marks around the centre "as if caused by a ligature, for the purpose of holding them with a willow or some such bandage".

By the end of the century the existence of these tools and early workings on Copa Hill and their probable pre-Roman origin were well reported within the pages of mining pamphlets, such as in the promotions and descriptions of Cardiganshire mines (Liscombe & Co. 1869-1870; Spargo 1870), as well as in the classical historical treatise on British Mining (Hunt 1884, 38). All these appear to have used Smyth or Francis as their source. It seems strange therefore, that no mention of stone mining hammers at Cwmystwyth is to be found amongst the many archaeological and antiquarian reports which frequently refer to the occurrence of similar tools at other mid-Wales mines (Arch.Camb. 1859;1860;1861; Williams 1866; Evans 1897,234). A problem then, as today, of archaeologists' poor familiarity with mining and geological literature.

A more recent source of information on early mining, particularly at Cwmystwyth, is to be found in Professor O.T.Jones' classic study of the lead-zinc mining district of Cardiganshire and Montgomeryshire. Within the latter the location of the ancient workings and the occurrences of stone hammers on Copa Hill were briefly described and illustrated (Jones 1922, Plate 7).

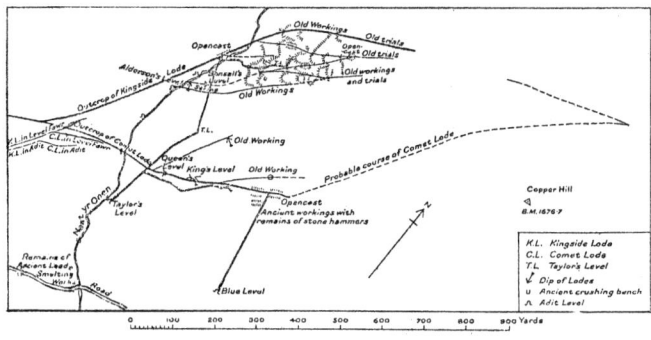

Figure 7: part of O.T.Jones' 1922 sketch map of Copa Hill with location of stone tools and ancient smelting site marked

FORMER ARCHAEOLOGICAL WORK

In 1936 Oliver Davies carried out a field survey of early mine sites in mid and North Wales (Davies 1935-1936, mss; 1936) selecting a number of these with surviving evidence of stone tools, charcoal or accompanying slag for purposes of future excavation (Parys Mountain, Great Orme, Nantyreira, Newtown Park, Cwmystwyth), in most cases simple trenching and sampling of associated spoil mounds. During the Spring or Summer of 1937 he carried out work at Cwmystwyth, cutting three stepped half trenches, each of about 4 to 5 metres long, into the sides of the South-East and Central Tips downslope and the Lateral Tip adjacent to the Comet Lode Opencast on Copa Hill (Davies 1937). From within these tips numerous examples of stone hammers and 'hammer querns' were recovered, lying in a stratigraphy of 'stone chips, black peat, brown earth and clay', but amongst which he claims to have found 'practically no charcoal' (Davies 1947, 59). A pre-spoil tip land surface consisting of a thin dark grey clay 'with charcoal points' and a light grey clay etc. was examined at the base of the section through the Central Tip, the opinion of the soil scientist Mr.E.Roberts of Bangor University being that this represented the bleached horizon of a podsolised soil which must have developed under peat, the latter having being removed ('cut') before mining operations commenced, most probably during the 'Roman era' (Davies 1947, 61). No other artefacts or dateable materials were recovered from the excavations, thus Davies did not elaborate any further on the date of these operations, although he did correctly made the distinction between the early forms of stone tools or 'hammer querns' found here and the larger stone anvils or 'cup-marked querns' found elsewhere (such as within the Kingside Lode workings on Copa Hill and below the Graig Fawr), most of which seemed to have been used in conjunction with iron hammers at a later date for ore crushing. He was also aware of the unique occurrence of stone hammers, confined to this one location at the top of the copper-bearing Comet Lode, although the absence or apparent absence here of copper ore within the tips then led him to think that the excavations upon this part of the vein were for lead, a conclusion which can only have helped confirm the belief that this was a Roman or Romano-British mine.

Despite the indications of antiquity, the site received little further attention until the start of the present investigations in 1986, the only other inspections for which there is any record being those carried out by Simon Hughes (Hughes 1981, 43-44) who also claims to have found hammer stones lower down on the Comet Lode outcrop, and Alan Tyler, who was able to re-identify one of Davies' trenches and further describe some of the stone tools which he had referred to (Tyler 1982). Tyler questioned the evidence provided by Davies for exploitation during the Romano-British period, noting Davies' own doubts over the extent and age of podsolisation which had taken place on the ground surface beneath the tips prior to mining. Nevertheless, he stopped short of speculating further on the age of the opencast, recommending only that the sort of archaeological work then being carried out by the Dolaucothi Research Committee should be extended to here, as well as to other sites in Wales including Parys Mountain, the Great Orme and Draethen (Tyler 1982, 315).

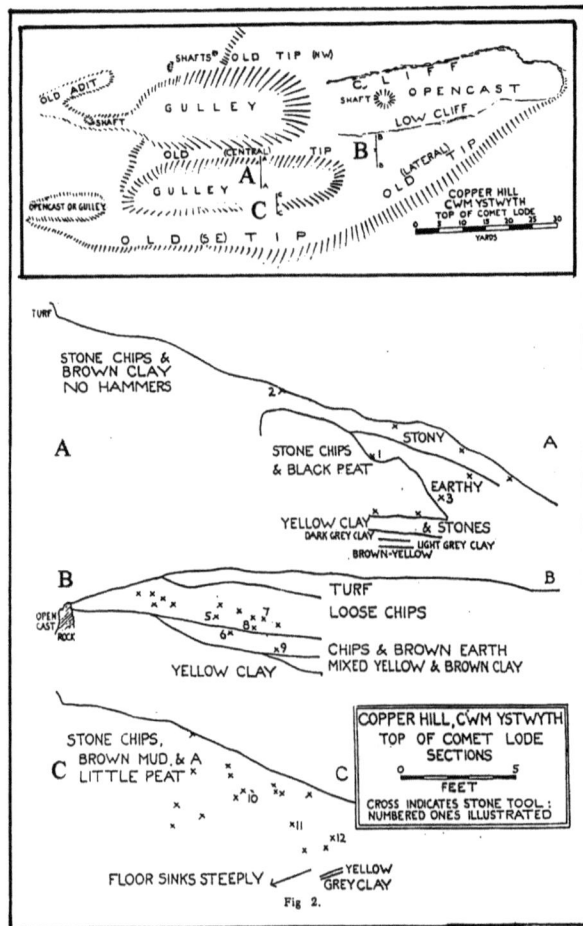

Figure 8: O.Davies' 1937 map and sketch sections of his excavated trenches cut through the tips of the Comet Lode Opencast (with position of hammers [x] marked (after Davies 1947)

CHAPTER 4

LANDSCAPE SURVEY AND ARCHAEOLOGICAL INVESTIGATIONS ON COPA HILL 1986-1999)

LANDSCAPE SURVEY

The complexity and richness of detail of the mining landscape visible on the north side of the Ystwyth Valley, particularly upon Copa Hill, was such that it was decided in 1989 that comprehensive archaeological recording and field walking of the whole area was necessary in advance of any major programme of excavation within the Comet Lode Opencast (Timberlake 1990b). For example, at least ten historic phases of mining and prospection have been identified upon these slopes (Timberlake & Mighall, 1992, 40), many of which are overlain features of a more recent agricultural landscape (peat cuttings, peat-drying platforms, sledding tracks, cultivation beds, field walls and *haffottai*), within which some earlier medieval-postmedieval elements (such as platform houses and trackways) and the traces of a former prehistoric landscape (including a cairn and cist, quartz boulder cairns, and the prehistoric opencast mine) are still recognisable. Understanding this sequence of land-use and exploitation has in fact been essential to the process of recognition of the early mining evidence, and indeed the reasons why it may have been confined to this location.

At the time of the re-investigation of the prehistoric opencast in 1986, an area of approx. 20 000 square metres which lay within the vicinity of these workings was surveyed by the author with the help of the National Archaeological Survey (RCAHM Wales), using fixed EDM points and a chain, tape and compass survey (Timberlake 1987,18). This survey area was considerably extended during the summer of 1989 to include the whole of the western and south-western slopes of Copa Hill, and part of the area bordering this to the west (eg Pant Morcell and Penguelan), an area of approx 1 square kilometre (Timberlake & Mighall 1992, 39). Meanwhile a more detailed contour plan was constructed by tape and level of the opencast prior to excavation (Timberlake 1990b, 8), whilst in 1994 an EDM survey was carried out by Dyfed Archaeological Trust of the extensive prehistoric tips below this. Further detailed EDM planning was undertaken by the Early Mines Research Group in June 2000 of an area of approx. 100 square metres lying to the east of the 1986 survey.

The results of all this survey work might best be summarised through a catalogue or description of each mining or agricultural monument or group of landscape features, their frequency and location, and the approximate period(s) to which they belong (SEE Fig 9; p.12). In addition, archaeological sampling/evaluation trenches or pits were cut at a number of different locations as part of a broader archaeological assessment of Copa Hill, and the results of this work are reported here in brief.

Prehistoric

MINING

Early Bronze Age Opencast [SN 81167523]

Figure 9: Comet Lode Opencast before excavation (D.B.Hague 1986)

Cutting in rock approx. 45 m long and 20 m wide (south end), and 8 m (north end) aligned SSW-NNE on vein with change of direction to SW-NE at north end. Present ground surface (top of infill) at 426 m at south end, 440 m north end, with buried profile of cutting possibly up to 10 m below this, but probably much shallower, perhaps only 2-3 m deeper at northernmost point. Exposed north-west cliff face of working considerably higher (6-7 m max.) than on south-east side (1-2 m), although both of these sides slope inwards towards centre of the buried working (which is thus much narrower). However, several small overhangs are still visible above ground surface along part of the north-west face (the hanging wall of the main vein). Within the middle of the opencast at ground height lies a mass of recent slate scree fallen from sides, part concealing the cone-shaped collapse of a modern prospection shaft and its accompanying small spoil tip located within the centre front of the hollow, and the floor is uneven throughout. The amount of weathering and scree-spalling of the rock walls has almost obliterated any signs of original working present upon the exposed rock face, except for that along a prominent fault, the course of which bisects the middle of the hanging wall, and also in the north-west corner. No marks of stone tool use, nor of firesetting, or any evidence for mineralisation (apart from quartz) can be seen upon the rock surface above ground level, and no stone tools were found within the opencast itself.

Figure 10: Location map (Cwmystwyth) and survey of Copa Hill (EMRG 1989)

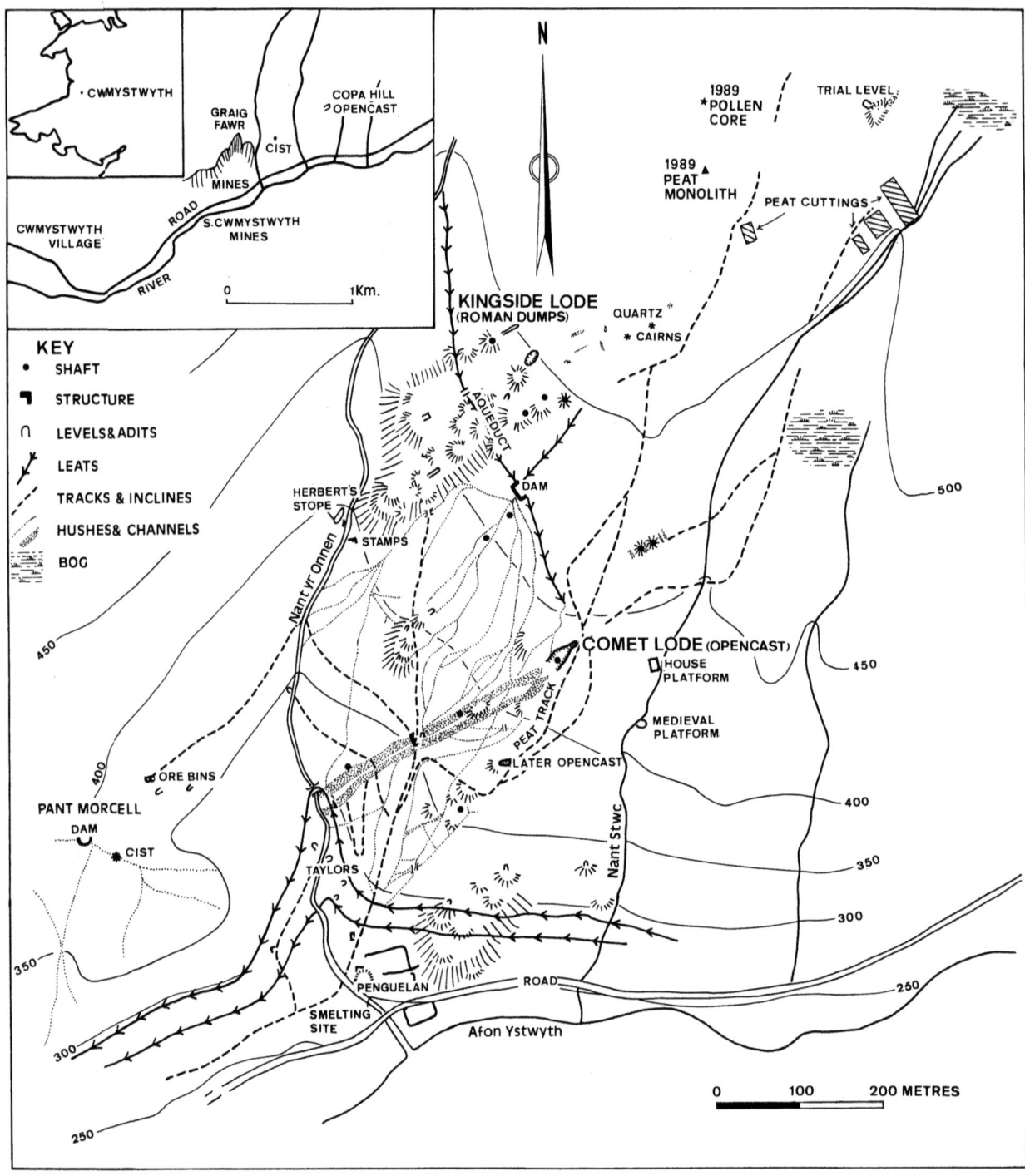

Bronze Age Spoil Tips [81137520 (Central Tip); 81177523 (Lateral Tip)]

Figure 11: View from downslope of tips dissected by hush gullies (ST 1993)

Consisting of four low (5-20') angle and denuded, elongate spoil mounds disguised by a thickly developed turf horizon only along the uneroded south-east side (Lateral and Upper portion of the South-east Tip) and within the lower portion of the largest Central Tip (110m x 20 m max.). Elsewhere, as a result of water erosion (gullying), and more recently over-grazing, some 35% of the tip surface now lies exposed. Most of the spoil was originally tipped downslope of the lip of the Comet Lode Opencast, possibly over previously worked portions of the vein, although much of this must have been carried a considerable distance downslope by hand. Where exposed this spoil occurs in layers, as well as in a mixed-up context consisting of broken pieces of slate and mineral vein in a matrix of small crushed rock flakes, gravel sized quartz, sand, a brown-yellow soil, and ubiquitous broken (now lichen-covered) hammerstones. Over the years many of these have eroded out, rolling downhill to collect within the base of the gullies which lie in between the tips, although occasionally some can be found hundreds of metres further downslope. Visible charcoal is rare, almost non-existent at surface, although faint traces of fire-reddening and bleaching are detectable on some of the fractured pieces of slate, along with mineral staining from copper (green), iron (brown), lead (pink and white) and manganese (black).

Drainage channel from the prehistoric mine [SN 8112751 to 80947512]

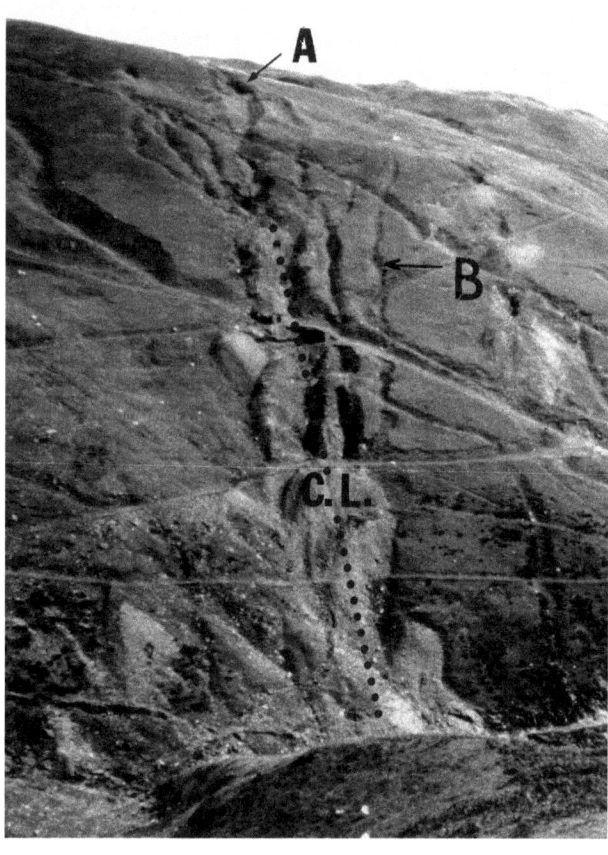

Figure 12: Hushing on Copa Hill showing prehistoric(?) drainage channel (B), course of Comet Lode, and opencast (A)

One of the several flood channels or gullies which follow the course of the outcrop of the Comet Lode downhill and which has the appearance of being a hush. This feature first appears as a shallow ditch (approx. 1m wide) at the top of the slope just below the small sloping plateau to the south-east of the prehistoric tips and mine, and from there on it can be followed downslope for a 200 m forming a wider and ever deeper channel until the point where it is abruptly terminated by the southernmost of the two later hushing channels which mark the outcrop of the lode itself. The course of this early channel is also readily visible on the 1:7500 Ordnance Survey 1973 vertical air photo of Copa Hill.

Some 15 m from its top another smaller channel branches off to the north-west, and from this, 10 m further down, several short parallel off-set trenches (2-3 m long by 0.5 m deep) appear to have been cut, at right angles to the direction of slope. These are suggestive of a secondary function for the channel, either in prospection (which seems doubtful) or perhaps for ore-washing and separation within artificially constructed pools. Unfortunately, there appears to be no exact parallel for this feature at other prehistoric mines, although ore-washing sites which may be prehistoric are known to exist on the Great Orme (Lewis, A. 1996) whilst P.R.Lewis (1977, 34) refers to a slightly more sophisticated arrangement of washing tables which he calls the 'goldwashing cascade' below Melin y Milwyr at the Dolaucothi Roman Gold Mine near Pumpsaint. There is a more recent reference to the use of the "Golden Fleece" method at the above site (Cauuet et al. 2000); one which may have used moss and brushwood barriers within pools or sluices for the entrapment of gold or other dense ore particles. Davies (1947, 63) also speculated upon the use of a similar system of washing tables here on Copa Hill, but never found the evidence to support this. If any such mineral separation took place this would probably have been for the removal of copper from a mixture of lead, zinc and copper ores, and at present there is insufficient evidence available to develop this interpretation further.

The dating of this channel system to the period of the exploitation of the prehistoric mine is mainly linked to the cutting of a number of exploration trenches during 1986 (D1a, A1, C1), 1989 (C2, C3, & C4), and 1999 (A3), which proved the continuation of this up-slope beneath a mantle of hillwash deposits and then underneath the prehistoric spoil mounds to the lip of a large rock-cut drainage/entrance trench situated at the front south-eastern corner of the opencast. The profile of this buried channel, and its infill in places with mine debris and ore fragments, is suggestive of occasional flood conditions, such as might have resulted from the breaching of the flooded opencast following an attempt to re-work it after a period of temporary abandonment. More tantalising is the possibility that this was an example of prehistoric hushing executed to uncover the course of the vein downslope, although other than its similarity with the copious examples of later hushes upon this hillside there really isn't the evidence at present to support this. However, the channel could still have served as a drain or soak-away for the mine, perhaps even as a washing channel, and if so, was probably used over a much longer period of time.

BURIAL

Bronze Age cairn and robbed cist, Pant Morcell [SN 80627502]

The remains of what is probably a damaged and almost obliterated cairn with a possible cist hollow has been identified upon the top of flat plateau-like summit of Pant Morcell, a prominent rock spur which projects out from the north side of the Ystwyth Valley at a height of 386 m OD. The cairn may have been damaged as a consequence of hushing during the late 18th century, when a small hushing pond was constructed less than 50 m above it, and a v-shaped channel was dug through its northern half. Quite possibly the site was levelled in advance of this, whilst cairn material could also have been robbed during the construction of the hushing dam. Indeed, what seems to be a small robber trench could be the result of spurious excavation by 18th century miners or others, as does the presence of 5 to 6 large and now scattered rock slabs which may once have formed the orthostats, base, and capstone of a cist. The circular outline of this 20 m diameter cairn, which was probably originally no more than 0.5 m high, can still be made out, and its position, constructed around a shallow rock outcrop which forms the summit, dominates the valley at this point. The cairn is overlooked by the Bronze Age mine some 600 m to the east on the opposite side of the Nant yr onnen valley, and situated some 40 m above this.

The lack of any evidence for smaller mound material in between the larger displaced slabs may suggest that this was originally a kerb circle, examples of which are common within the Plynlimon area, as at Hirnant and elsewhere (Leighton 1984), although if this was the case, the degree of damage present has destroyed all traces of the orthostat kerb.

Figure 13: Pant Morcell cairn (photo ST 1991)

Bronze Age burial cairns are not common in the landscape in and around the Upper Ystwyth Valley, there being only 9 recorded examples within an area of 100 square kilometres circumference of the mine, the nearest of these being more than 3 km away to the south at Domen Milwyn (Ceredigion SMR; Briggs 1994).

Quartz boulder cairns (SN 81247560 and 81257561)

Two small cairns (3 m diameter and < 0.5 m high) composed almost exclusively of white quartz boulders and approx. 30 m apart were located on a low ridge overlooking a small peat hollow approx 100 m north-east of the top of the Kingside Lode workings on the west slope of Copa Hill. The possible site of a third cairn 30 m further east along the same ridge is marked by a depression with several quartz boulders on the edge, perhaps indicating extensive robbing for white quartz which now marks the route of an old trackway leading up to peat cuttings on the moorland above. None of the quartz used in the construction of these cairns seems to have been derived from the extensive quartz-rich mine spoil which lies a short distance to the west, suggesting that these probably pre-date any exploitation of minerals on this part of the hillside. It would seem that these are rather small for burials, yet the lack of any other cairns within the vicinity which might indicate later agricultural clearance, their ridge-top location and exclusive use of quartz boulder material, is much more suggestive of prehistoric construction.

Medieval

MINING

Ancient leat and aqueduct (SN 81047577 to 81157523)

Figure 14: View NW along line of leat (ST 1986)

The course of an ancient leat which once fed a small hushing pond located within the infilled prehistoric opencast has been traced over a distance of 500 m (with a vertical drop of about 40 m) from its source within a small embayment at the head of the Nant yr onnen stream at 483 m OD, to a point about 50 m above the opencast.

From here a rapid drop (presumably once controlled by a sluice) must have fed water down to a rock-cut channel still visible high on the north wall of the opencast and into a reservoir below. An overflow or release channel branches off downhill some 20 m before this enters the opencast.

The relative antiquity of this leat compared to the later examples on Copa Hill is evident from the degree of infill and erosion and collapse of the downhill side turf embankment and stone revetment Moreover, this clearly pre-dates the 18th century hushing dam (under which it is buried) as well as the majority of the older workings on the Kingside Lode which seem to have obliterated its course in several places. Within this latter section the remains of a narrow drystone-walled aqueduct which must once have carried the leat in a wooden launder around a small cliff-face can be seen at SN 81087550, whilst elsewhere the leat was constructed to be about 1-1.5 m wide and about 0.5 m deep, therefore narrower and also much shallower than the 19th century examples lower down.

In 1993 the shallow leat infill was archaeologically examined at 3 different locations (F3-F5). In places waterlain shale gravel and silt was still visible on the floor of the channel, variously covered with pockets of peat and/or clay, and commonly sealed by a stone tumble from the revetted embankment. From one of these sections (F5) cut close to the source of the leat a sample of this buried peat was removed from a depth of about 20 cm, which has since provided the following C14 date: OxA-10041, 789±33: 1190-1290 cal AD (Stuiver et al. 1998; OxCal v3.5 Bronk Ramsey 2000). The probable interpretation therefore is that the leat is Early Medieval in age; last being used sometime before 1300 AD to supply water to the infilled prehistoric opencast for the purposes of hushing to recover lead on the Comet Lode (most likely during the period of lease or working of the mine by the monks of Strata Florida). Unfortunately, the above date refers only to the final silting up and abandonment of this channel, and as such we cannot completely exclude an earlier Roman or even pre-Roman date for its use, a caution suggested by the the medieval dating of the peat-filled Annel (Gwenlais) leat at Dolaucothi (Burnham et al. 1992). However, there is now other (dating) evidence from within the opencast itself (SEE this vol.) which points to a medieval phase of hushing, and by inference therefore, a similar date for the use of the leat. A late 15th-century leat of similar dimensions (the Lumburn Leat) has also been described from the area of the Bere Ferrers silver mines, Devon, though used for water-powered pumping rather than hushing (Claughton 1996).

Hushes on the Comet Lode (SN81147521 to 80877506)

These two parallel hushes, the more northerly of which marks the principal outcrop of the Comet Lode, are amongst the deepest and most prominent of the many hushing gullies on Copa Hill (SEE Fig.11). Extending over a strike length of some 300 metres from the mouth of the prehistoric opencast to the floor of the Nant yr onnen, these appear to have been excavated by repeated hushings from a pond or series of ponds situated within the infilled and levelled opencast, the main channel exiting from a breach in a dam wall located between the ridges of the Central and North-West (prehistoric) tips, whilst another channel must have exited at almost 90' to this a few metres to the east. Some 150 m downslope of the opencast the sides of the main hush are up to 3 m high, cut through an overburden of natural moraine and solifluction deposit, whilst the exposed vein, in places from 0.5 to 1.5 m wide, has been worked by means of shallow opencasting as well as by more modern shafts, stopes and adits.

The archaeological evidence, as discussed, proposes that this hushing originated in medieval or pre-medieval times (a Roman date cannot be ruled out), although some of the field evidence suggests that these hushes were further deepened within their lower parts during the later 18th century, as a result of an intentional or accidental contribution from the more southerly release channels which radiate out from the still intact hushing dam high up on the western slopes of Copa Hill (SN 81137542). The amount of rock and soil overburden released and transported by hushing from Medieval times onwards may have significantly contributed to the silting up of the Ystwyth River, as evidenced from the accumulated fan of detritus which has built up downstream of the mouth of the Nant yr onnen.

Early adit working on the Kingside Lode (SN 81057553)

Although known as the 'Roman Dumps' (Davies 1947, 58), perhaps on account of the apparent antiquity of the numerous collapsed shallow adit workings and abundance of anvils or hand mortar stones amongst the tips, there is actually very little proof today for any sort of medieval or pre-medieval mining. However, the evidence for this may lie beneath much later workings, which have in total produced an estimated 200 000 tons of waste. One small clue may be found in the form of a small square-cut trial adit, no more than 1 x 1.5 m long, the cutting to which appears to have been bridged by a stone aqueduct constructed to carry the early leat, and thus is presumably an earlier feature. Indeed, many of the older lead workings may lie on the edge of the moorland above, where considerable numbers of costeaning pits are to be found.

Lead roasting and smelting hearths, Penguelan [SN 80907484]

Recent investigations in 1996 and in 1999 of the site of an 'ancient lead smelting works' (Jones 1922, pl.VII) located upon a small plateau-like apron of moraine at the foot of Copa Hill, just to the west of the Nant yr onnen and above the road, has yielded fragmentary evidence for a number of destroyed smelting hearths (probably bole furnaces) and roasting beds. Fieldwalking identified an area of infertile grassland and some strewn lead smelting debris, but without any evidence for up-standing structures. However, a rapid geochemical soils survey carried out in 1996 using portable XRF equipment (Spectrace 9000) revealed the probable sources of a number of major lead anomalies (Jenkins & Timberlake 1997), whilst geophysics, including magnetometer and electromagnetics surveys carried out in 1999 (by Archaeophysica Ltd.) helped to identify the likely sites for roasting or burning along the tops of three shallow ridges. Sample trenches (2m x 2m square) were then cut in three different places (PSS 1-3), two of which revealed traces of damaged and/or dismantled hearths at between 25 and 50 cms depth, one consisting of a 10 cm thick bed of charcoal raked out over a fire-reddened ground surface, the other the base of a collapsed stone structure within a matrix of poorly fired white clay. Samples of slag, fused furnace lining, litharge, and runnels of lead metal were present within the overlying debris layers. A sample of oak charcoal provided the following C14 date: Beta-140992, 820±60: 1030-1300 Cal AD (Stuiver et al. 1998) - a result which may suggest the smelting of lead here by or on behalf of the Cistercian monks of Strata Florida Abbey.

AGRICULTURAL & DWELLING

Platform house and fieldbanks [SN 81297522]

Figure 16: Platform House (ruinous), Nant Stwc (ST 2000)

The stone platform and ruinous foundations for a combined dwelling and animal shelter, possibly a medieval or early post-medieval *lluest* has been identified some 100 m east of the prehistoric opencast by the side of the Nant Stwc. The rectangular shaped platform with (17 m x 7.5 m) with rounded corners has been constructed of large slabs of stone lain directly onto a former surface of grass and rushes, surrounded by a drainage gully, and with foundations for 1 m thick side and gable walls, plus a central partition wall and hearth between the animal shelter and single room of the dwelling. Most of these walls now survive as a shallow tumble, since robbed for stone, whilst traces of other smaller structures plus several small field banks can be seen a little distance to the south. The building seems to conform to some of the descriptions of *lluest* (summer dairy (James 2001, 64), or

Figure 15: Penguelan Lead Smelting Site; **TOP LEFT**: Sites PSS1 and PSS2 (in background); **BOTTOM:** collapsed stone and clay-bonded furnace structure in trench PSS3; **RIGHT**: Magnetic susceptibility survey showing likely position of furnace remains (photos ST)

alternatively may have been a dwelling place for shepherds (*haffotai*); but is altogether different in plan from the 18th -19th century smallholdings present upon the lower slopes.

Unidentified platform site [SN 81277517]

A small stream-side platform, consisting of a semi-circular flat area approx. 2m x 1 m wide cut into the top of the exposed rock ridge above the east side of the Nant Stwc gorge, was examined at the point where this meets the top of the steep northern side of the Ystwyth Valley. The rim of the platform here is surrounded by a small orthostatic kerb on its downslope side, whilst its floor has been built of from layers of broken shale and soil, which towards the top contains some burnt rock and charcoal. Although a prehistoric origin for this structure seemed possible, a sample of oak charcoal instead returned an early medieval date: OxA-10045, 704 ±34: 1250-1390 Cal AD (Stuiver et al. 1998). A number of other less well defined platforms were surveyed within an area of 100 m to the north of this, and to the east of the prehistoric opencast. It is not clear whether they are related in any way.

Early Postmedieval

MINING

Within 100 m radius to the south and north of the Comet Lode opencast are to be found a number of pre-19th century mining remains. These include the Comet Adit, a small wet stope and level of unknown age located only 25 m to the north-west of the opencast; and downslope of the prehistoric spoil tips, within the base of the northernmost hush gully on the Comet Lode, is the 'Copper Level' (Timberlake 1994 - Russ Soc), a probable mid-18th century powder and pick-cut level, from the end of which a rise had been driven upwards some 30 metres to connect with the base of the prehistoric working, presumably to sample the unexploited vein (SEE Fig. 4b). The original entrance to this level appears to have been obliterated by the excavation to surface of the stope workings from King's and Queen's Adits lower down on Copa Hill (post-1911). However, a little over 100 m to the south-west of the prehistoric opencast is to be found another small primitive-looking opencut, barely 17 m long, which is located at the top of a minor lead-zinc vein which courses parallel to the Comet [SN 81107512]. A collapsed adit connects with the base of this, and it seems possible therefore that these are contemporary, and perhaps also of a similar age to some of the earliest workings on the Kingside Lode - therefore 16th-early 18th century. However, a medieval date is still possible, although a prehistoric origin seems very unlikely, Davies (1947) having sampled the tips here in 1935, finding no trace of associated hammer stones (Davies 1947, 58). Some 100 m north-west of the prehistoric mine on the moorland edge, a short series of shafts, spoil mounds and prospection trenches follow the continuation of the Comet Lode. These appear to be trials, and typologically are rather similar to the late 17th-early 18th century

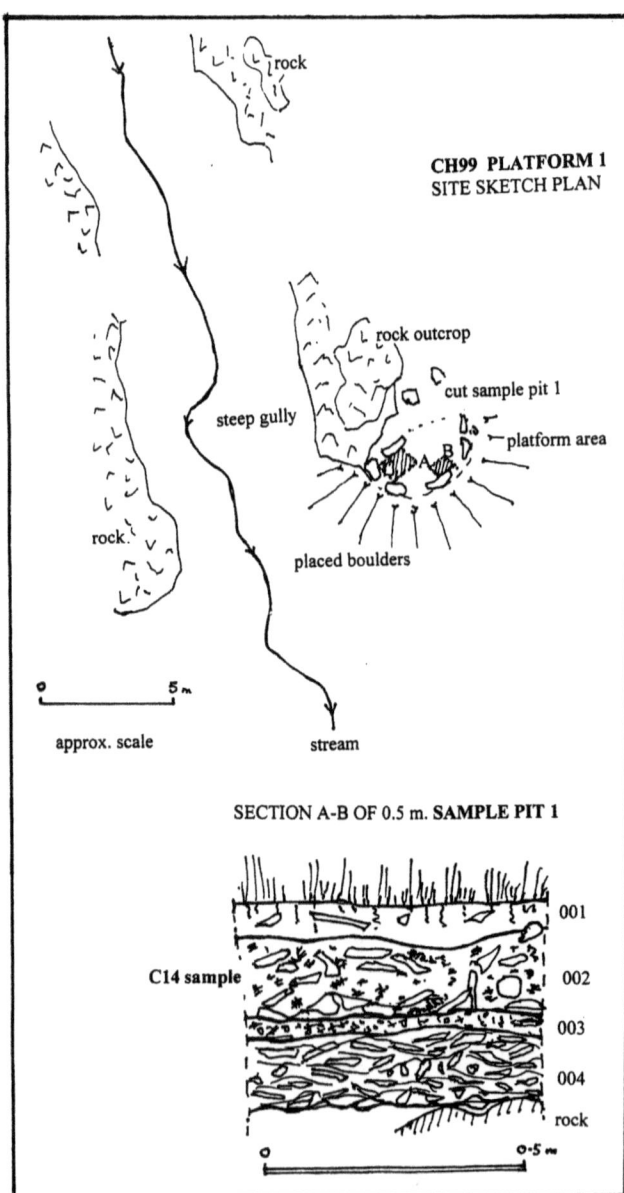

Figure 17: Plan and section of Nant Stwc Early Medieval Platform

workings carried out by William Waller at Pen Trefach Mine (Hughes 1981). The buried outcast from these trials was thoroughly sampled in 1993 for any traces of prehistoric mining activity, the evidence for which may have lain hidden beneath the peat, and then have been disturbed by the more recent mining. None was found.

The largest complex of mine workings on Copa Hill are undoubtedly those already referred to as the 'Roman Dumps'. These workings exploited several branches of the Kingside Lode, and consist of at least 15 extant yet collapsed adits with associated opencuts, and possibly others buried beneath successive layers of spoil. Some of these galleries are believed to extend more than '300 yards from surface' (Davies 1947) and are associated with a whole series of accompanying hand-dressing floors and the tumbled down remains of drystone-walled dressing sheds (Armfield 1989), some of which contain stone-lined square buddles (approx. 0.5 m diam.) for washing and separating lead ore, as well as crushing benches associated with stone anvils or mortar stones. Some of the latter possess single and others double cup-shaped mortar depressions, whilst many have been worked on opposite faces until the point of fracture. Indeed the majority of these mortar stones could have been used in conjunction with iron hammers right up until the end of the 18th century, perhaps even as late as 1803, at which date most of the ore on Copa Hill was still being crushed by hand (Davies 1947, 58). The inclusion of broken anvil stones as well as rock fragments with the impressions of early hand-drilled shot holes in within the make-up of the walls of some of these dressing sheds supports the view that their use continued up to and beyond the introduction of gunpowder circa. 1750, although it is quite conceivable that some of the dressing sheds, if not the mortar stones within them could date from the late 16th century, or before (Timberlake & Mighall 1992). For instance, the large opening of Herbert's Stope, on the opposite side of the stream, dates from the 1670's, whilst we know that William Waller's (and later Bonsall's) water-powered stamp mill located at the foot of the 'Roman Dumps' [SN 80907535] was there in 1699 (Hughes 1981,63). We can only presume therefore that some of the upper (now lost) workings, such as Bayard's Adit, may have been Elizabethan or earlier. Rather similar, but slightly later 18th century workings complete with square buddles and mortar stones are to be found overlooking the road on the south side of Copa Hill [SN 81047491], just to the east and north of Blue Level, an adit driven before 1847 to intersect with the Comet Lode some distance beneath the prehistoric opencast (Smyth 1847).

The majority of the visible hushing remains on Copa Hill emanate from the probable late-18th century hushing pond and dam [SN 81137543], perhaps constructed during the Bonsall era (Hughes 1981,16). These hushes therefore avoid and post-date most of the workings on the Kingside Lode, as well as some of those on the Comet, although an intermediate series of workings lying between them [at SN 81007525] appears to have been discovered in the process (Timberlake & Mighall 1992, 40). The earlier Comet Lode hushes were dammed up to allow several of the release channels to cross over to the south, whilst in places prospection trenches were cut [as at SN 81017503] perpendicular to these, in order to prove the course of any minor veins intersected. It seems perfectly possible therefore that up to 25 other prospection trenches (both short and long forms) found during the course of this survey on Copa Hill may date from a similar late 18th-century period, including those within the vicinity of the Comet Lode Opencast. Indeed, finds of clay pipe stems encountered during the sectioning one of these trenches (CH99 P3) some 50 m south of the prehistoric mine, would tend to support an 18th or early 19th-century date for their excavation.

Figure 18: Remains of dry-stone walled shed with anvil stone(s), Kingside Tips (ST 1992)

Figure 19: Probable 18th-century hushing dam, Copa Hill (ST 1992)

A miner's track, presumably hundreds of years old, ascends Copa Hill from the foot of the Nant yr onnen valley up to the 'Roman Dumps'. The first part of this follows the course of the later incline up to King's Adit, but thereafter it is preserved in its entirety, with a raised embankment downslope, perhaps to facilitate the passage of carts or pack-animals [SN 81007513 - 81207538].

Victorian and Modern

MINING

Many of the recognisable early nineteenth century workings are to be found lower down on the south and west slopes of Copa Hill, and these include deeper shaft workings (Penguelan Shaft) as well as a succession of shafts and levels (Burrells Adits) and longer drivages (Penguelan Level, Blue Level, Bonsall's Level, and Alderson's Level). Somewhat later came the construction of the stone bank and bridges for the self-acting incline (300 m long and 2 m wide) that climbs from the Nant yr onnen to meet up with the two tramroads from Aldersons and Herbert's Adit workings on the Kingside Lode, and which dates from the period of John Taylor and the introduction of centralised ore-milling. Probably contemporary with this are the associated ore bins [at SN 80987513] and a square building(s) with thick walls of partly faced mine rock cemented with a distinctive course lime mortar, as well as the lower leat channels which can be seen following the 280 m and 292 m contours, and which supplied water to turn waterwheels on the main dressing floors to the west of Copa Hill. A wide brick-arched adit portal, now half-concealed beneath the bed-load of the stream, marks the site of Taylor's Level [SN 80887502], the principal late-19th century drivage under Copa Hill which worked both the Kingside and Comet Lode portions of the mine.

Twentieth century mining on Copa Hill appears mostly to have been confined to the Comet Lode. The fairly wide entrances to the King's and Queen's Adits, plus various structures around these, and the spoil tips which follow the contours beneath each of the tramways, belong to the period 1911-1914, whilst trials made by the Gallois company in the 1930's to look for the continuation of this vein, include the remains of drill stems, a wooden rig, and the site of a short adit, and these can be seen high up on the moorland plateau north of the summit [SN 81657589]. The remains of 1m x 1m square wooden lined assay pits or shafts [e.g. at SN 81047550] can be seen upon the flattened tops of many of the old tips within the area of the 'Roman Dumps', and date from 1923, when they were excavated to sampling the lead/zinc content of these with a view to re-working the old spoil (Hughes 1981, 42).

AGRICULTURAL

Smallholding and fields at Penguelan [SN 80947489]

The poorly preserved remains of a single dwelling, of which all but a small fragment of the north and west gable wall has disappeared into a large depression formed by subsidence at the mouth of the nearby Penguelan Mine Adit, is to be found associated with 3 small fields and fieldbanks upon the plateau above the valley road, immediately to the east of the Nant yr onnen stream. 'Lazy-bed' strips, probably originally for the cultivation of potatoes, can be seen within two of these enclosures, the best preserved example being that on the east, with fifteen 10 m x 1 m wide beds which follow the dip of the slope. 'Penguelan' appears to have been inhabited during the first half of the 19th century, but had probably been abandoned by 1911, at which date the mouth of the adit had already caved-in. This may well have been the same dwelling known as 'Nantyronnen' in 1728.

Peat-sledding track [SN 80977506 - 81267557], **drying platforms** [SN 81137516], **and peat cuttings** [e.g. SN 81537573]

These and other examples form several bits of evidence for small-scale peat extraction on the top of Copa Hill (Timberlake 1990 a; Briggs 1991; Timberlake & Mighall 1992, 43), a rural subsistence-related activity presumably once common throughout the Welsh uplands during the postmedieval period (Owen 1975), but which nevertheless today continues to remain a quite under-researched subject.

Along the edge of a small plateau some 60-100 m south-west of the prehistoric opencast are to be found the remains of some 5 or 6 rectangular platforms, of which at least 4 lie in pairs, facing downslope, with traces of stone foundations and loose slabs at their lower, and slightly raised ends. The lay-out of these structures conforms tolerably well with various historical descriptions, both of the methods of stacking and drying peat in Cardiganshire and Merioneth (Owen 1969), and to eye-witness accounts, such as that given by Edward Pugh in 1816 of rows of drying stacks visible along the sky-line tops of the scarp slopes below moorland turbaries at Dinas Mawddwy (IN Owen 1975, 316-7). On Copa Hill the position of these structures, close to the start of the steep sled route, with their narrow walled ends facing the prevailing wind, would have been ideal for the purposes of stacking of peat.

The rectangular depressions of the old turbaries or peat cuttings can be seen within the two peat basins north of the summit, the largest within a small peat-filled valley at the head of the Nant Stwc. From here, several bifurcating tracks or turbary roads can be followed over the moorland, the route in places marked by white quartz boulders, to a point where these join above the prehistoric opencast and then begins to descend more steeply as a well-worn sled track which zig-zag's its way downwards to the floor of the Nant yr onnen valley.

Just below the opencast parts of this track are fairly deeply rutted, and sections have been by-passed by several alternative branches, some of which give the appearance of being hush channels rather than tracks. However, the separation of these rut marks conforms well with the 3 foot (0.9 m) width of the wooden peat sledges described by Owen (1975).

Furthermore, the use of a metal-detector in 1989 to survey the areas where these tracks had cut through or else disturbed the prehistoric spoil mounds, revealed a quantity of ironwork including a number of small horseshoes, nails, chain, and part of a sled runner. This evidence is fully consistent with descriptions of the use of ponies and iron-shod sleds in more recent times (Owen 1975), and it also suggests that peat extraction here must have continued well into the 19th century.

Figure 20a: Plan of peat sledding tracks, turbary stacks (peat-drying platforms), and hushing channels within vicinity of Comet Lode Opencast (ST and N.A.S.(RCAHMWales) 1986)

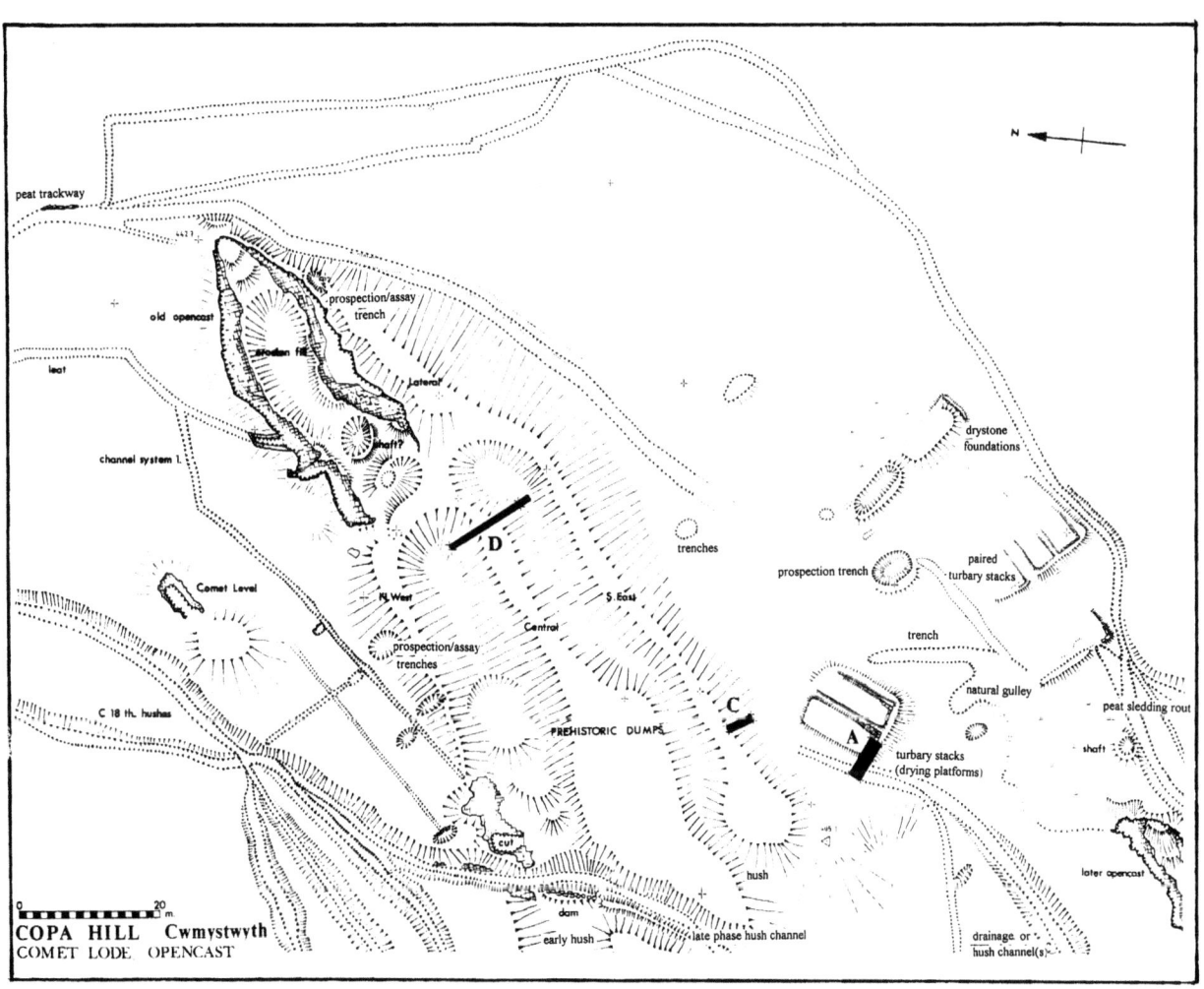

Figure 20b: Contour plan of environs of Comet Lode Opencast showing hushing channels, tracks, tips and all excavation trenches cut by EMRG (1986-1999). Compiled from survey data of EMRG and from surveys commissioned by EMRG from NAS(W) and DAT. Drawn by B.Craddock.

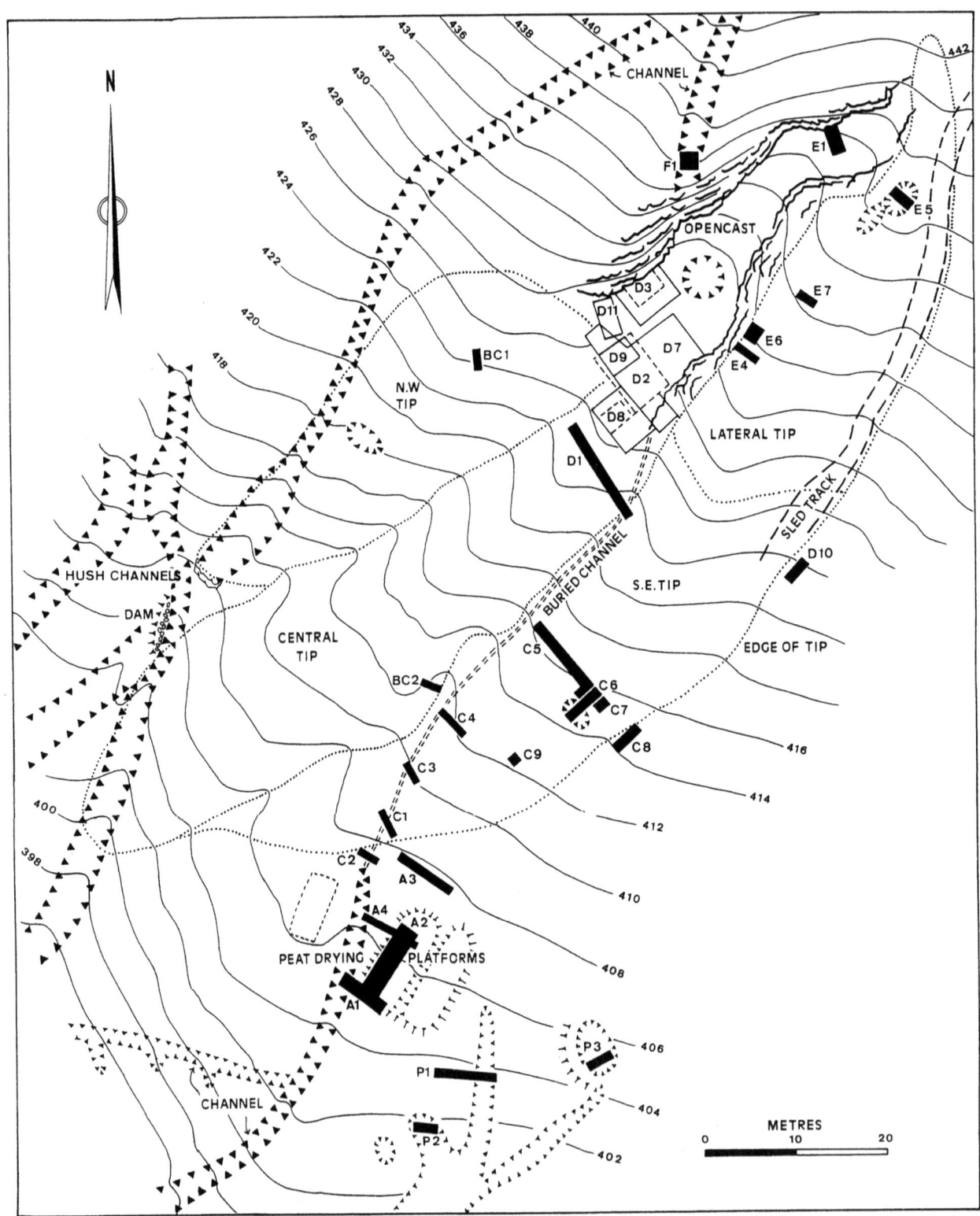

CHAPTER 5

EXCAVATIONS WITHIN THE VICINITY OF THE COMET LODE OPENCAST (1986)

The main objective of these investigations carried out by the author (with the support of the National Museum of Wales) in September 1986 was to re-sample the tips first examined by Davies for the purposes of obtaining material suitable for radiometric dating. Three evaluation trenches were undertaken.

Trench D1

The principal excavation consisted of a 12 m long and 1 m wide trench dug by hand across the head of the main spoil tip (Central Dump) immediately downslope of the prehistoric opencast (in fact this was sectioned some 10 m below its mouth). This revealed at the very base of the spoil profile the outcrop of a shallow south-westerly sloping rock surface, the latter encountered at depths of between 20 cm - 200 cm below respective north and south ends of the trench, and giving the appearance of being a natural weathered ground surface rather than that of an original quarried and since buried opencut. Only at the extreme southern end of this trench was there any evidence for an artificial cut through the underlying land surface (here consisting of silt and clay layers [D1a 006a+b] lying within the base of a natural trough located between the Central and SE Dumps), this appearing to be a drainage channel or hush initiated before or else during the main phase of tip deposition.

Figure 21: View looking NW along trench D1 through Central Tip showing old ground surface with buried channel in foreground (photo ST 1986)

The estimated depth of this tip close to its crest (between 2 and 2.5 m) is useful in that it provides us with an indication of the maximum thickness for the spoil mounds surrounding the mine (over an area of 2400 square metres), a large proportion of which must be less than 0.5 m thick.

The detail of the tip layers in section revealed a complex sequence of inter-digitating lenses of coarse and finely broken rock, crushed vein material and quartz, and finely crushed (and now largely oxidised) sulphides with larger lumps of discarded galena, washed rock fines, shale gravel and clay, and layers rich in charcoal. Within some parts of the tip these layers have become heavily cemented by iron/manganese pan, and the stratigraphic details obscured. The latter had formed a number thick red/brown ochre (iron) and sooty-black (manganese) bands, suggesting the position of former perched water table(s), with zones of reduction beneath, and in places waterlogged with fairly good organic preservation.

Four main layers or phases of tipping were detectable, the succession and direction of tipping throughout being from north to south, a fact which may indicate that mining was then being undertaken somewhere to the north-west of the dump, perhaps against the north wall of the opencast, from which point rock debris was being removed from the mine to be tipped progressively further downslope, forming a large apron of spoil to the south-west. However, a hiatus in deposition, followed by either a short period of erosion and undercutting, or else an artificial re-cut of the south-eastern slope of this tip, separates layers 4 and 3 (Fig. 22). Here the overlying layer (3) consists of a markedly different sort of material from that which forms the voidy layers of coarse and finely broken rock (4) beneath, the former possessing a much greater proportion of finely crushed mineral, burnt rock, ash and clay, perhaps reflecting a renewed phase of productive activity. This horizon, which splits up southwards into a number of sub-layers or lenses, some of which consist mostly of sterile broken rock, others which are rich in charcoal (3e), and still others with a higher proportion of crushed mineral, vein rock and hammer-stones amongst them, today forms the bulk of the existing tip (this calculation excludes the volume of the surface layers 1 and 2 since lost to erosion). There is evidence here that the miners have exercised some control over the rate of tipping and the angle of slope of the spoil. This can be seen in the placing of larger boulders of vein rock over the bedrock surface, perhaps to help stabilise the base and to prevent the slumping of spoil downslope, or back into the channel or drain from the mine. This precaution may also help explain the re-cut of the tip slope. However, no evidence for a wooden revetment (along the base of the tip) was picked up in section. Excavation across an east-west step through the edge of the earliest tip phase (004 in Step 1), somewhat

Figure 22 a: Trench D1[east section] cut through Central Spoil Tip (ST 1986)
Figure 22 b: Photomontage of same section showing position of hammer-stone and antler finds (photo ST 1986)
Figure 22 c: Fragmentary remains of copper-stained antler discarded in mine spoil (ST 1986)

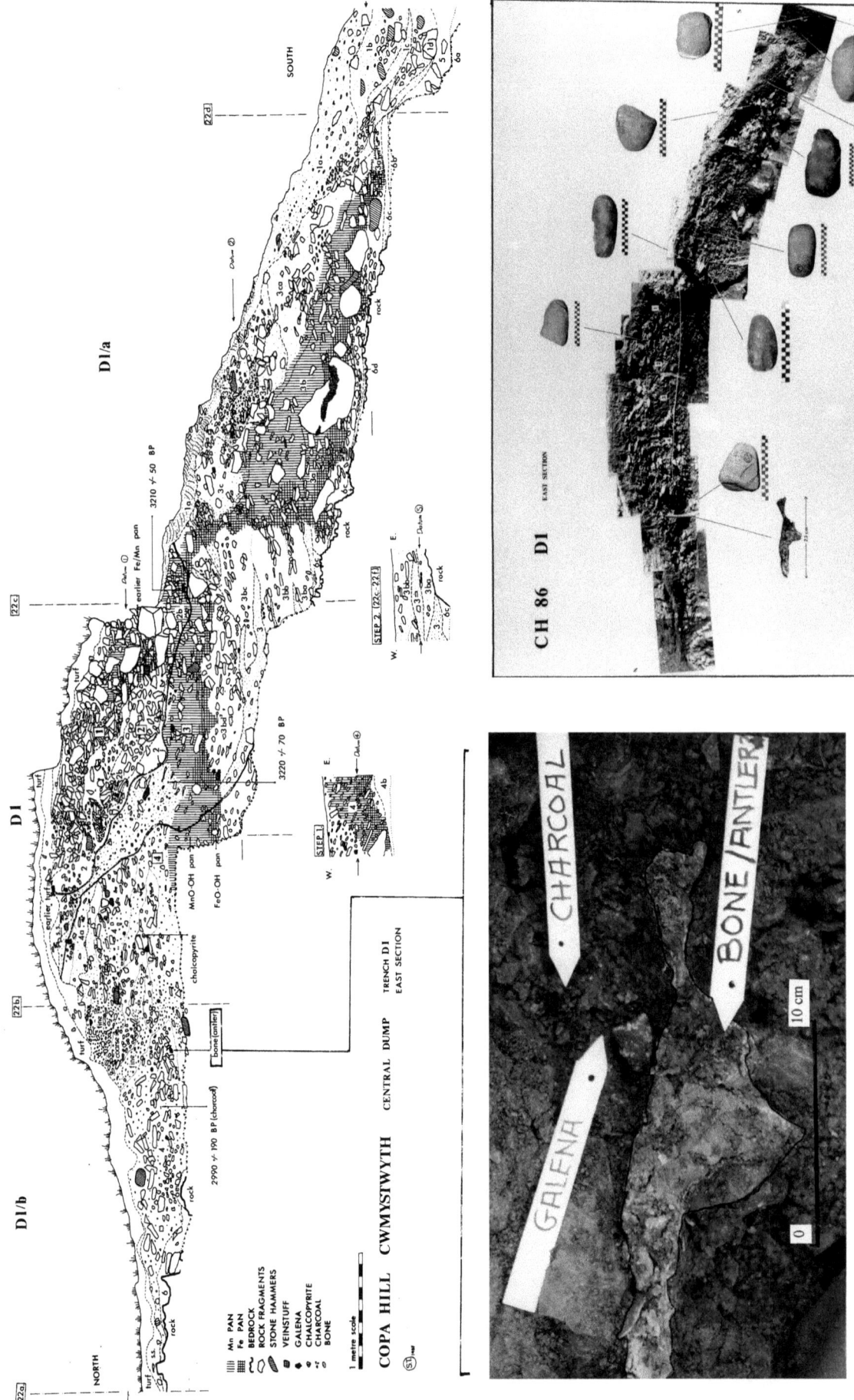

surprisingly revealed an easterly tipping direction, suggesting that the first tip, at least for a while, accumulated in an up-slope direction, again suggesting careful control of tip stability.

Altogether 87 discarded and broken hammer-stones (mostly of between 1-2 kg in weight) and spalls (the latter produced as a consequence of the fracture of their tips during mining) were recovered during the course of excavation of this trench, with more than 90% of these from layers 1-3, including those from the erosion deposits within the channel. Other finds included some unidentifiable carbonised wood (3e) and the rather decomposed and crushed remains of a damaged tine of deer antler (*Cervus elaphus*), the latter stained blue from copper salts, and found inside of a preservation halo present due to slightly higher pH surrounding a void in 004 (Fig. 22c). Associated with the latter were a few fragments of cobbed chalcopyrite, galena, and a hand-held crushing stone (Timberlake & Switsur 1988).

The above excavation produced the first radiocarbon dates for the site in 1987. These were taken from bulk samples of charcoal, reported originally to have been composed mainly of ash (*Fraxinus*), with minor hazel (*Corylus*), and oak (*Quercus)* (IN Timberlale & Switsur 1988), but since re-interpreted as being predominantly of oak branch and mature wood (Grant & Chambers pers. comm. in Timberlake & Mighall 1992, 42), collected from amongst fire-setting debris taken from each of the three main layers. The original calibrated dates are as follows:

D1 002 - Q-3078, 3210±50: 1590-1410 cal BC

D1 003a - Q-3076, 3220±70: 1685-1370 cal BC

D1b 004 - Q-3077, 2990±190: 1685-810 cal BC   (Stuiver & Pearson 1986)

The above dates (Timberlake 1987; Timberlake & Switsur 1988) now seem to be a little late (Early-Middle Bronze Age) for the operation of the mine, but the significance of this will be discussed later.

Trench C1

The apparently shallower SE Tip was sectioned towards its lower end, in order to try and establish its depth, its composition, and also to ascertain whether the spoil here was *in situ* or else had been washed down from above during the course of hushing. Over most of the area of this trench (4 m x 1 m) the buried ground surface was reached at a depth of only 10 to 50 cm, although

the presence within the lowest unweathered layer (003) of both small and large hammer-stone fragments, crushed mineral, quartz sand and clay (although admittedly little charcoal), makes it seem likely that a thin layer of spoil had already formed here during the Bronze Age, even if most, if not all of this had been derived the large spoil tip(s) immediately below the opencast. From this location rock debris may have been washed downslope over the space of a couple of hundred years. The relatively early deposition of spoil at this location is also supported by the lack of humus development or the presence of any significant scree beneath the tip, as well as by the discovery of a channel buried beneath it (possibly the same channel as that encountered 50 m slope within Trench D1). The greater part of this seems to have been infilled contemporaneously with a waterlain shale gravel containing both hammer-stones and re-worked dump material (004 + 004b), and then conformably overlain by the spoil tip, apparently without any really significant break in time. The irregular cross-section and longitudinal profile of the channel suggested periodically fast-flowing stream conditions, the position of the trench coinciding with a drop in its floor from 0.5 to 1 m depth, thereby providing an informative section through the old clay land surface at this point. Within this section, some 30-40 cm below the pre-tip ground surface (a dark clay and rootlet horizon = 005) a 5 cm thick clay layer was noticed with numerous flecks of charcoal, a layer which may well reflect the earliest human activity on site. This overlay an orange clay, silt and shale, elsewhere recognized to form the natural ground surface above the moraine.

Figure 23: Looking south across top of excavated and emptied channel beneath spoil of SE Tip (Trench C1: photo ST 1986)

Trench A1

A shallow trench was cut adjacent to the south-west edge of the westernmost peat-drying platform in order to prove the existence of the southward continuation of the same channel picked up beneath Trench C1 (the traces of which were only just discernable at the top of the steep scarp slope). This revealed only a 25 cm deep peat-filled depression, and this feature was not excavated further. However, the turf and soil cover overlying a 6m x 2m area of wall tumble at the south end of the associated platform was removed and the stonework underneath inspected and planned in an unsuccessful attempt to try and understand the function and age of this (then) unrecognised structure. However, amongst all this, and clearly re-used here as building material, was found a large double-sided saddle-quern type mortar stone (Fig. 24b), a grinding implement which appears originally to have been used in conjunction with a stone rubber or maul, and therefore quite distinct from the stone anvils previously described from the Kingside 'Roman Dumps'. A prehistoric origin for this quern now seems likely, associated with primitive ore crushing and processing activities carried out somewhere below the opencast; yet the exact whereabouts of its original location remains a mystery. Perhaps this was re-discovered by the peat-workers half-embedded within one of the tips (a context also noted at the the EBA tin mine at Kestel in Turkey (Willies 1990)), or perhaps under some half-buried structure since robbed for stone.

Figure 24 a: Stone tumble of drystone-walled foundations of peat-drying stack with turf removed. The re-used saddle-quern type ore mortar is shown as found (ST 1986)

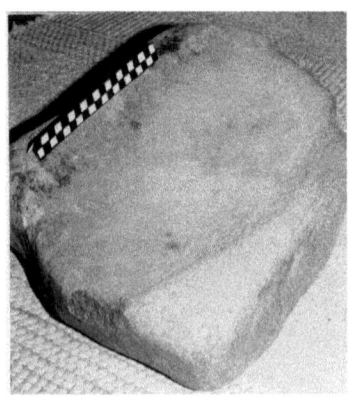

Figure 24 b: 'Saddle-quern type' ore mortar stone. Detail of one side (ST 1987)

Figure 25: Section across buried channel in Trench C1 (ST 1986)

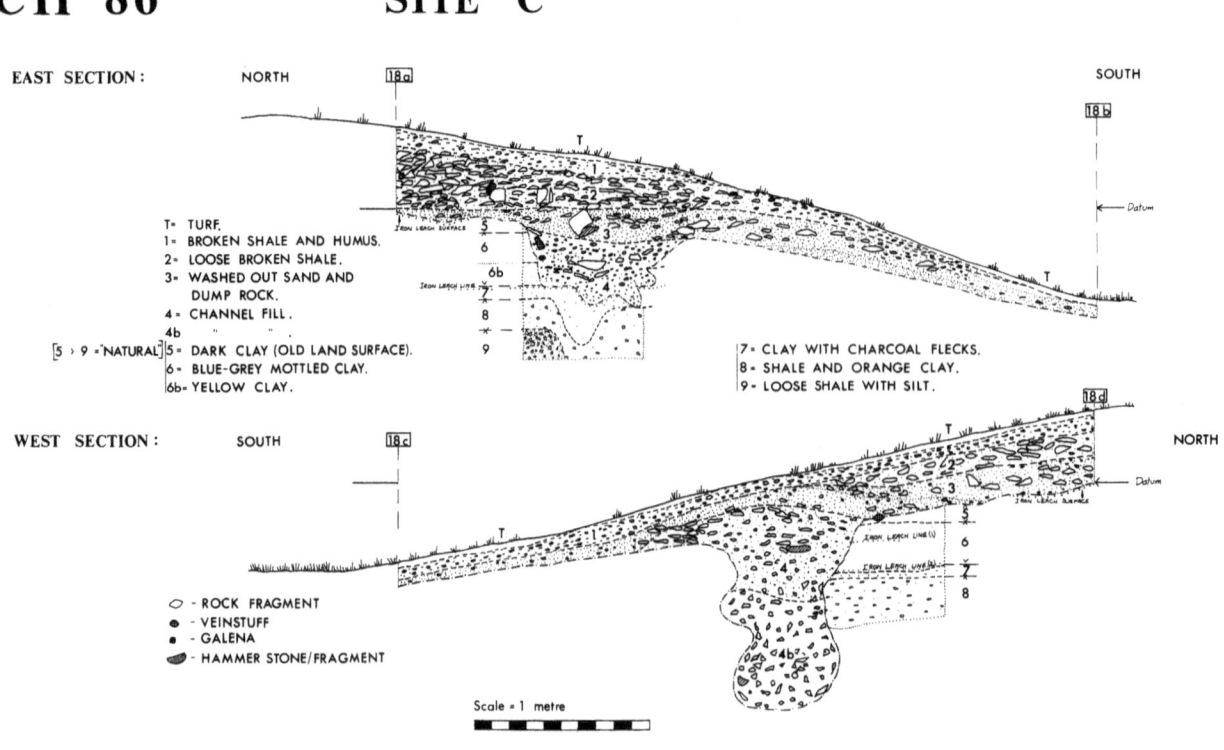

CHAPTER 6

EXCAVATIONS WITHIN THE PREHISTORIC MINE 1989-1999

1989

Following the confirmation of Bronze Age mining a major programme of research excavations was undertaken from 1989 onwards by the newly formed Early Mines Research Group (with the support of the National Museum of Wales), its specific purpose being to investigate what appeared to be a deeply infilled and undisturbed prehistoric working. Two main areas were chosen for excavation - one at the foot of the 3-4 m high cliff face on the north side of the opencast, where an initial 2 m x 2 m trench was sunk to investigate the depth of deposit (later increased to 25 square metres - Area D3), and one 6.5 m x 5.5 m cut (Areas D2 - D5) which lay above the plateau-like front of the opencut, and beneath which it was believed a shallower sequence of sediments covered the original entrance(s) to the mine. Both of these sites were excavated by hand, all areas except D3 being backfilled in August 1989.

Area D2

The hummocky relief here suggested more recent disturbance of this area following the infill of the mining hollow behind, a fact confirmed by the discovery of a shale and turf bank (004), subsequently recognized as the remains of an old hushing dam built across the front of the opencast. Behind this appeared an accumulation of light grey/brown clay, stones and silt (005), being interpreted as a hill-wash deposit derived from outside the working up-slope to the north-east, upon the surface of which (just beneath the turf) were found a number of re-deposited stone artefacts, including a hammer-stone and four 'stone lids' (Fig. 96E). Below and behind the dam lay a 15-20 cm thick silt (009), whilst at the base of this a more stony horizon lay upon what one might suppose to have once been the clay floor of the hushing pond, a sticky clay layer which sealed the bottom of the turf-stack dam wall, and which overlay what seems to have been a much earlier sub-soil and shale scree (010), a waterlain sequence of stratified silts, clays, sand and shale (011), plus the tip of an organic silt (013) resting upon turf-covered prehistoric mine spoil (015). The latter appeared to be a thin spill of mine debris which had slumped back into this south-eastern corner of the working from amongst surrounding piles of tip during a comparatively early stage in its abandonment, and this was separated from a rather more substantial layer of mine spoil containing both charcoal and stone tool fragments (017) by another layer of organic silt (016 & 027) - the latter with recognisable branch wood (*Betula sp*), sphagnum moss, and beetle remains, and found lying against one of the rock faces within a small sondage at a depth of 1.5 m. A still deeper sequence of *in situ* mine tip (030-034), some of it with suggestions of layering formed from alternate horizons of crushed mineral, charcoal, and loose mined rock, was exposed within the base of 2.5 m deep east-west trench (D4) cut as an extension to D2, the latter for the purpose of sectioning the wall of the hushing dam. A radiocarbon date for this was obtained from one of the enclosed charcoal layers produced by firesetting. The latter confirmed that Early Bronze Age mining had indeed produced this sealed and buried mine spoil [BM-27320, 3500±50: 1964-1694 cal BC, which rounded out to 1970-1690 cal BC (Pearson & Stuiver 1986)].

The turf-stack dam at the top of this section has since been identified as a probable post-medieval structure on the basis of another radiocarbon determination of the thin peat layer or ground surface (032) which lies immediately beneath it [Wk-9543, 387±39: 1430-1640 cal AD (Stuiver et al. 1998)]. The latter rests upon several thin, laminated, and well indurated gleyed clays originally identified as possible mineral processing slimes (019-020), but conceivably the remains of an earlier hushing pond overlying embanked spoil at the entrance to the opencast, a suggestion reinforced by the presence of post-holes (023 & 037), some of which appear to pre-date the turf-stack dam, and which could therefore be retaining features for an earlier clay or shale bank. If such a phase of hushing existed, then it is most likely that this would be contemporary with ancient leat which enters the opencast behind, and therefore of Roman to Early Medieval date. Some re-processing of the tips and also washing of lead ore may also have taken place within the vicinity of the opencast during the latter period.

The best evidence yet for hushing activity within the opencast, apart from the turf dam (within which individual layers of turves can still be recognized), must be the excavated remains of a sluice channel with a large stone slab resting in its base. The latter was found at shallow depth overlooking the top of the hush gully which this helped to create - that lying between the Central and North-Western Tips. However, the supposed southern exit from the hushing pond (i.e. that which it is assumed created the gully between the Central and South-East Tips) could not be found within two further trenches (D5 & D6) cut to the south of D2.

Instead, the north-west to south-east turf dam appeared to thin out against a natural bank of existing hill-wash deposits which one can only presume formed the south wall of the 25 + sq metre reservoir. This increases the probability that the south channel was formed earlier, possibly during the time of working the prehistoric mine.

Figure 26: Remains of stone sluice obstruction in base of release channel from ? Medieval hushing dam (photo ST 1989)

Figure 27 a: South section of Area D2 showing turfstack dam [004] (B.Craddock 1990) TOP

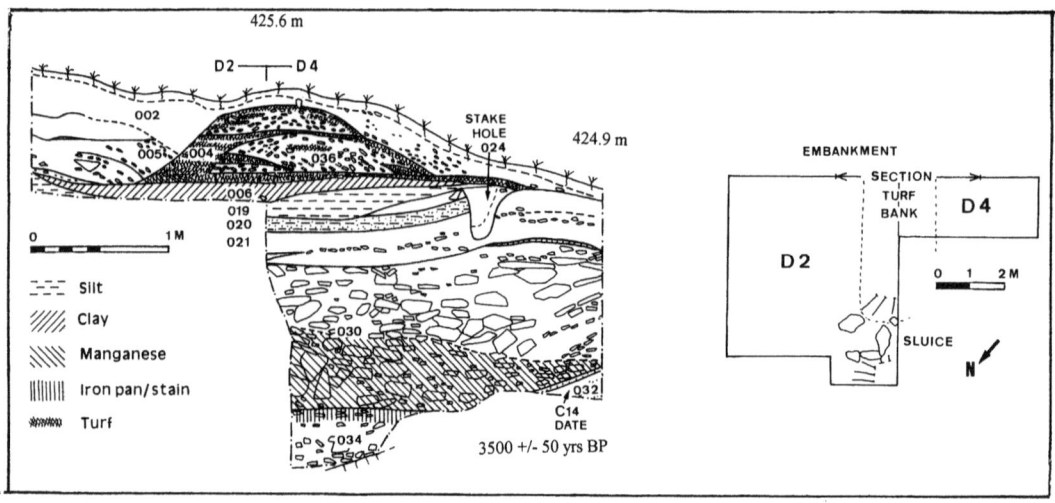

Figure 27 b: SE sondage D2 showing top edge of BA organic infill (B.Craddock 1989) BOTTOM

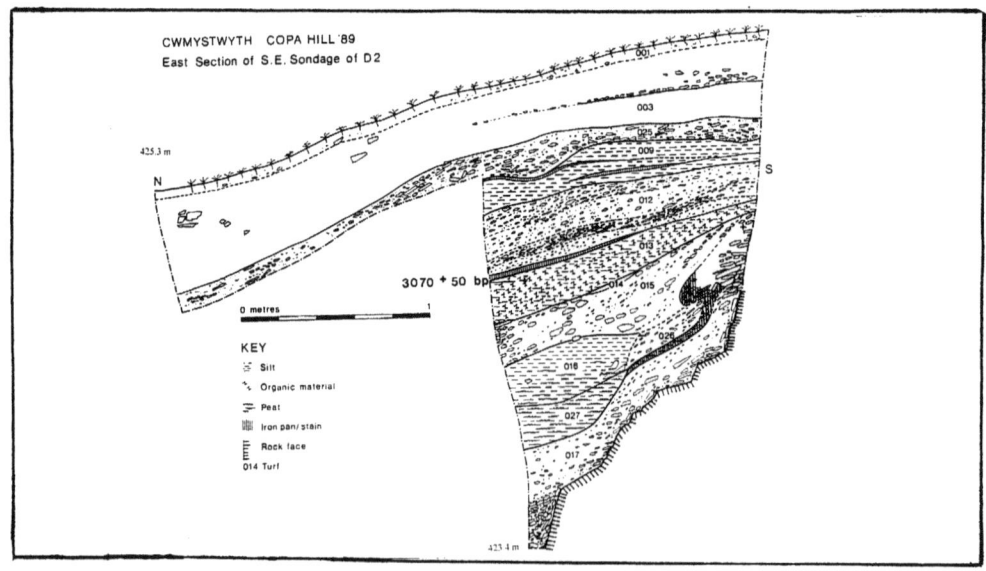

Figure 28: Exposure of prehistoric mine gallery beneath opencast cliff overhang, Area D3 (photo P.Craddock 1989)

Area D3

The location of this trench and a 16 sq m planned area was chosen beneath a large rock step at the foot of the north wall of the opencast. The site suggested a buried overhang, and if so, it seemed likely that this would have preserved early mining deposits at much shallower depths than elsewhere.

Found beneath some 20-30 cm of recently eroded shale scree lay the top of a shallow spoil tip which could be traced north-eastwards towards the cone-shaped collapse of an old shaft within the centre of the opencast. The spill of this contained large lumps of blocky mineral vein, some with abundant copper ore, and evidently derived from modern mining, the latter showing signs of blast fractures with several large 1.25 inch (2.3 cm) diameter shot holes (familiarity with the various different periods of mining at Cwmystwyth, plus some scant historical evidence for shot-hole sizes (Earl 1968), suggests that these are most likely to be late eighteenth or early nineteenth century). Removal of this spoil tip revealed the partially levelled pre-mining deposits of natural scree and orange sub-soil, the angle of which now dipped at some 15-20 ° towards the centre of the opencast, whilst the walls of the working itself began to steepen, and a small void appearing between this and the rest of the shale infill. At its widest point this void was followed downwards, a stepped trench being cut back from the cliff, the rock face of which began to develop a prominent overhang at a depth of about 1.5 m, though still without any signs of tool marks on the freshly broken frost-shattered exterior. However, further digging revealed the roof of a 'cave' opening some 2 m below surface, that excavation subsequently showed to be one side of an arched fire-set gallery, approx 0.8 m wide at top, 2 m wide at its base, 1 m high, and just over two metres long (at the deepest point excavated) - the latter exhibiting ample evidence for the use of stone tools in the form of blunt pitting marks left upon its roof and walls. The sculpting of the roof, in this case across the ends of the cleavage of the shale wall-rock, had left an undulating concave surface (Fig. 29 a + b), presumably following an attempt by the prehistoric miners to drive a short cross-cut from a swelling of rich ore upon one side of the opencut through a short distance of barren rock, to meet up with another known fault behind. The latter evidently proved to be sterile, and this short working was soon after abandoned. Overlying the top of peat deposits which had accumulated on the floor of this gallery were found a number of large blocks of rock which must have fallen many years afterwards, probably from the roof beneath the weathered entrance to this working.

Figure 29 a: Roof of mine gallery showing imprint of working using stone tools (chiefly the pounding action of mining hammers). Drawing by B.Craddock.
Figure 29 b: Photograph of detail of tool marks from same.
Figure 29 c (BOTTOM): Plan plus transverse and longitudinal profiles of top of mine gallery in 1989 (B.Craddock)

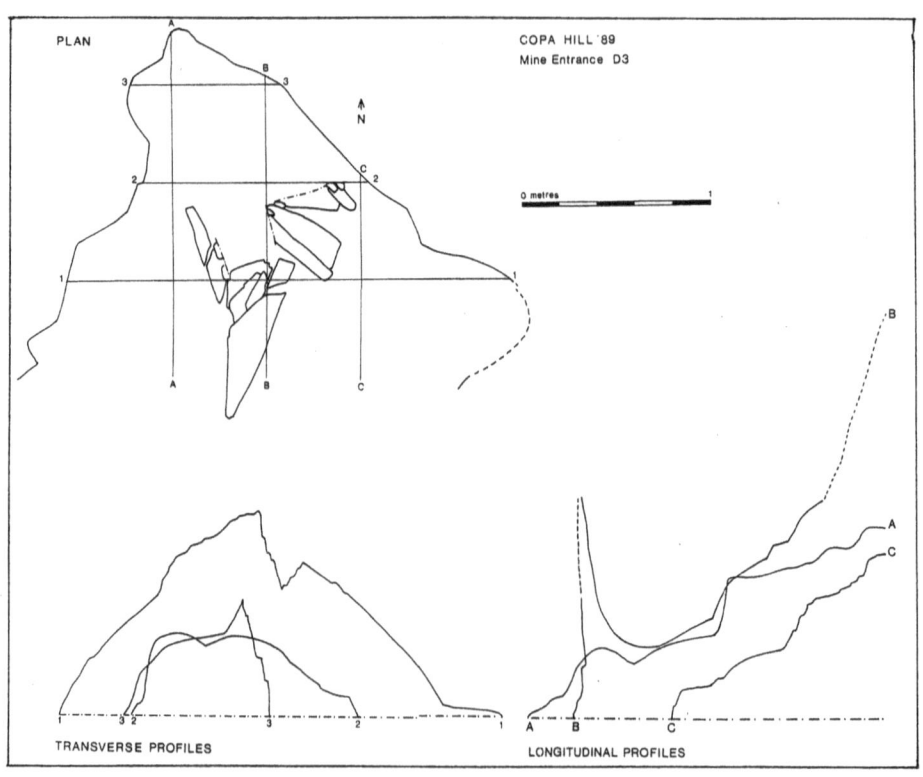

Figure 30: Recording within the part-excavated mine gallery in D3, 1990 (photo P.Craddock)

1990 - 1993 Deeper Stratigraphy - North side of opencast and Mine Gallery

Deeper excavations carried out in 1990 to uncover the floor deposits of the gallery meant that adequate shoring materials together with special procedures for underground working had to be brought together on site. The area of excavation was expanded to about 20 square metres, and a braced scaffold box-frame with tin and wood shoring behind it was bolted in place to a depth of more than 4 metres.

During the widening of the surface trench further artefacts were recovered from the horizon of modern mining spill associated with the visible shaft, including some fragments of late eighteenth or early nineteenth century dyed woolen fabric (pers. comm. P.Walton - Textile Research Associates), sawn and chopped bone food waste, wood chippings, and roof slates (Timberlake 1990), whilst the cutting of a proper section through the infill deposits beneath this revealed several layers of buried scree separated by lenses of laminated clay and silt (006/ 016). The latter possibly the same as those found lying almost directly below the level of the hushing pond in D2. Beneath this an earlier, steeply dipping sequence of weathered scree and soils lay banked up against the cliff face, and concealing the original entrance to the prehistoric gallery. In the middle of this sequence a dark turf layer (009), still surviving as a thin peat within the waterlogged base of the southern section, provided an important marker horizon for this large volume of natural infill, indicating a period of relative stability within the natural weathering/ scree accumulation cycle that allowed colonisation of these slopes by plants. An Early Medieval date was obtained from the C14 determination of this layer [BM-2780, 950±50: 990-1200 cal AD (Pearson & Stuiver 1986)]. However, a further 1.5 m of soil and scree deposit lay between this and the peat-covered floor of the gallery, further attesting to the age of the original working.

A much more complex abandonment history for this part of the site soon became evident following the fortuitous discovery of two additional intrusive mining features cutting the uppermost scree and soil layers within 5 metres of the rock face. A small shallow circular pit (106) was identified at the top of the south section cut into the surface of 006. Apart from its association with mine waste this would seem to be an unlikely candidate for a trial shaft. However, some 2 metres from the rock face was another feature (115), first detected as a small depression within the top of the same layer, which was soon to develop into a substantial (1 metre wide) sub-vertical round to square shaft, steeply inclined towards the rock face, and identifiable in plan view to a depth of some 3.5 to 4 metres. The same shaft appeared to cut the basal peat horizon (023) corresponding to that on the floor of the mine gallery, and then follow the footwall of the lode downwards another 1- 2 metres. At this point it was associated with considerable amounts of wood debris including split and broken oak staves which may have been derived from the collapse of the contemporary shaft lining, or perhaps from an earlier working still. Considerable iron-panning present within these waterlogged sediments, plus subsequent slumping of the deposits, had obscured much of the fine detail of this relationship. For instance, samples of roundwood collected from what appeared to be re-deposited or disturbed early mine spoil (031) close to the supposed bottom of this rapidly infilled later shaft produced an Early Medieval date [BM-2760, 830+/-140: 895-910 or 950 -1400 cal AD (Pearson & Stuiver 1986)], whilst less than 50 cm away, and from a very similar deposit, a much later date [BM-2828, 40+/-30: 1695-1735 or 1815-1925 cal AD (Pearson & Stuiver 1986)] was obtained from a discarded piece of sawn oak timber, the latter more credibly reflecting the date of this intrusive feature. It has been suggested for example that the medieval wood at this depth was re-deposited, derived from the collapse of the shaft sides run-in from above (Timberlake & Mighall 1992,38), but conceivably there may also have been another much earlier shaft in the vicinity.

Figure 31: Infilled shaft feature (light grey) in x-section shown cutting buried medieval soil horizon (dark band). Photo P.Craddock, 1990

Traces of this historic mining activity was also encountered in form of three small holes, presumably excavated with the tip of an iron pick, and located at a depth of between 5 and 5.5 metres below surface upon the sloping rock walls within the south-west corner of this trench, in an area otherwise covered with traces of stone tool work. Evidence for this shaft was lost at a depth of about 5.5 m within a disturbed layer of re-deposited prehistoric mine spoil (031), suggesting that these shallow trials were quickly abandoned and rapidly backfilled on the realisation that the earlier miners had effectively exhausted the vein.

Figure 32: Iron (?) pick hole on rock overhang to the side of the intrusive shaft sunk through early mining levels and mine infill layers. Hole approx. 4 cm across in middle (photo P.Craddock 1990)

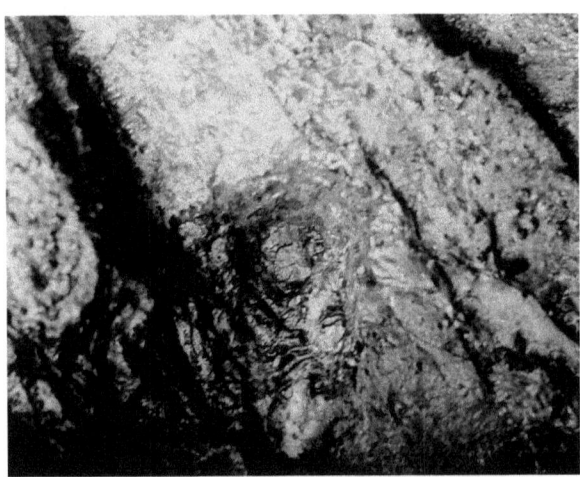

By contrast, the floor deposits present within the mine gallery were exceptionally well preserved. Sealed beneath the debris from roof collapse was a layer of fossil scree containing residual charcoal, most probably washed in from the roof and walls of the working, overlying a 10 - 20 cm thick layer of peat (019), the latter consisting of lenses of compressed leaves and branch wood (birch, hazel, oak etc.), mosses and grasses, some twig charcoal and shale fragments. Within this was uncovered the broken tip of a finely cut oak stake, the context and preservation of the latter suggesting an object washed or else thrown into the opencast, after little exposure to the external environment. Samples of this peat were taken for the purposes of pollen analysis, as well as for C14 dating - the latter returning a Late Bronze Age date [BM-2759, 2850+/-80: 1265-840 cal BC, rounded out to 1270-840 cal BC], clearly significant in that it provides us with a *terminus post quem* for the abandonment of this working, perhaps even of this whole area of opencast. In places this peat layer covered a mineral clay washed out of the faults, but in the centre of the floor of the gallery it rested upon shale (022), and beneath this a thin felty dark green peat (024). The latter was interpreted as a basal algal/grassy layer) found covering the mining deposits, here consisting of compressed crushed quartz sand, lamellar clay and charcoal (025) - effectively the final phase of mining activity which helped to create this gallery. However, careful excavation of what remained of this did not reveal any artefacts other than the three hammer-stones found together as a cache within the clay fill (020) of a worked-out hollow at its rear. One was a re-used flake tool (CH90:h15) complete with side notches for hafting (and so probably intended for use as a chisel/pick in such confined places); the others being a single re-worked flake recovered from the layer of crushed quartz (025), and a small spall from a hammer-stone (CH91:h9) found lodged within a hollow in the floor (032). The degree of compaction of this floor and the presence of crushed fragments of galena suggests that this ledge may once have been used as a platform by for ore crushing as well as other activities. This may have included ore washing, as suggested by the presence of several small interconnected hollows beneath the north-west vein that could once have held pools of water. The latter are still fed by an intermittent spring, from which a small channel now drains into the opencast.

Figure 33: Small cache of stone tools found *in situ* within worked-out hollow (photo P.Craddock 1990)

Figure 34: Same worked-out hollow (ribbon of ore) fully excavated showing stone tool (pick/chisel) marks and pool of water (spring ingress) at base with original drainage cutting (photo P.Craddock 1990)

Below the level of the mine gallery small lenses of sticky grey clay and ochre enclosing larger lumps of fresher-looking charcoal (029), and in some cases pieces of part-carbonised firewood (036), are all that remains of original firesetting deposits within this part of the mine. Only fragments of this deposit have survived the ravages of subsequent erosion, preserved within the cracks of ledges cut into the footwall of the lode. A sample of this charcoal (036) collected from a depth of about 4.5 metres, but perhaps coeval with that found upon the floor of the mine gallery, provided an Early Bronze Age radiocarbon date [BM-2812, 3460±50: 1915-1675 cal BC, which rounded out to 1920-1670 cal BC (calibr. Pearson & Stuiver 1986)].

The rock floor of the opencut, which in fact turned out only to be a ledge in the side of the working, was reached at a depth of 7.1 metres following excavations here in 1991. Incomplete running sections within the narrowing heavily shored trench revealed a continuation of the same peat layer as found within the mine gallery (019), now dipping steeply towards the centre of the opencast, and heavily broken up by post-depositional slumping. The latter overlying some 3 metres of almost vertically stacked angular scree (030) within which several thin interleaved layers of fine washed shale, clay and organic accumulation are to be found - some at least indicating breaks within the natural weathering cycle. At its base, a blocky shale layer (043) overlain by a thin spill of the same light grey clay and gravel layer containing crushed quartz and charcoal (031) appears likely to indicate the top of the re-deposited mine spoil. However, the lack of *in situ* Bronze Age mining horizons and the impracticality of continuing excavation at this depth resulted in the subsequent abandonment of this trench.

During machine-aided backfill of this excavation 1993, an opportunity was taken to widen the trench slightly when the shuttering was removed. This allowed examination of the southern side of the shaft cut, revealing part of a single oak mortice and tenon box-frame (SEE Fig. 35 & Fig. 36 [section E]) of the original metre square shaft lining - the latter used perhaps in conjunction with shuttering made up of split oak or hazel planks or hurdles, a technique formerly employed whilst sinking in soft ground. Rather similar constructions have been uncovered during excavations of 12th-13th century AD shafts associated with a silver mining settlement at Altenberg, Westphalia (Dahm et al.1998). Still other examples have been found within post-medieval mines, as well as in the Roman copper mines of Cyprus (Bruce et al. 1937), therefore these features are difficult to date on typological grounds alone.

Figure 35: Fragments of oak mortice and tenon square bracing from base of intrusive shaft (artefact drawings B.Craddock; reconstruction by ST)

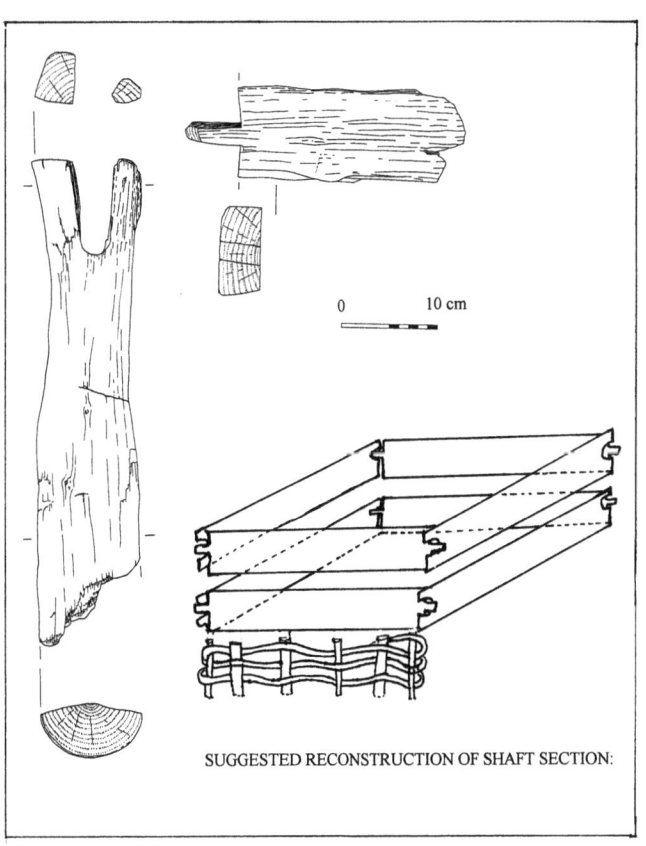

SUGGESTED RECONSTRUCTION OF SHAFT SECTION:

Figure 36: Sections (A-E) within Area D3 (cut between 1990-1993). Position as shown in location diagram in bottom RH corner. NB section C is upon floor of mine gallery (drawings B.Craddock)

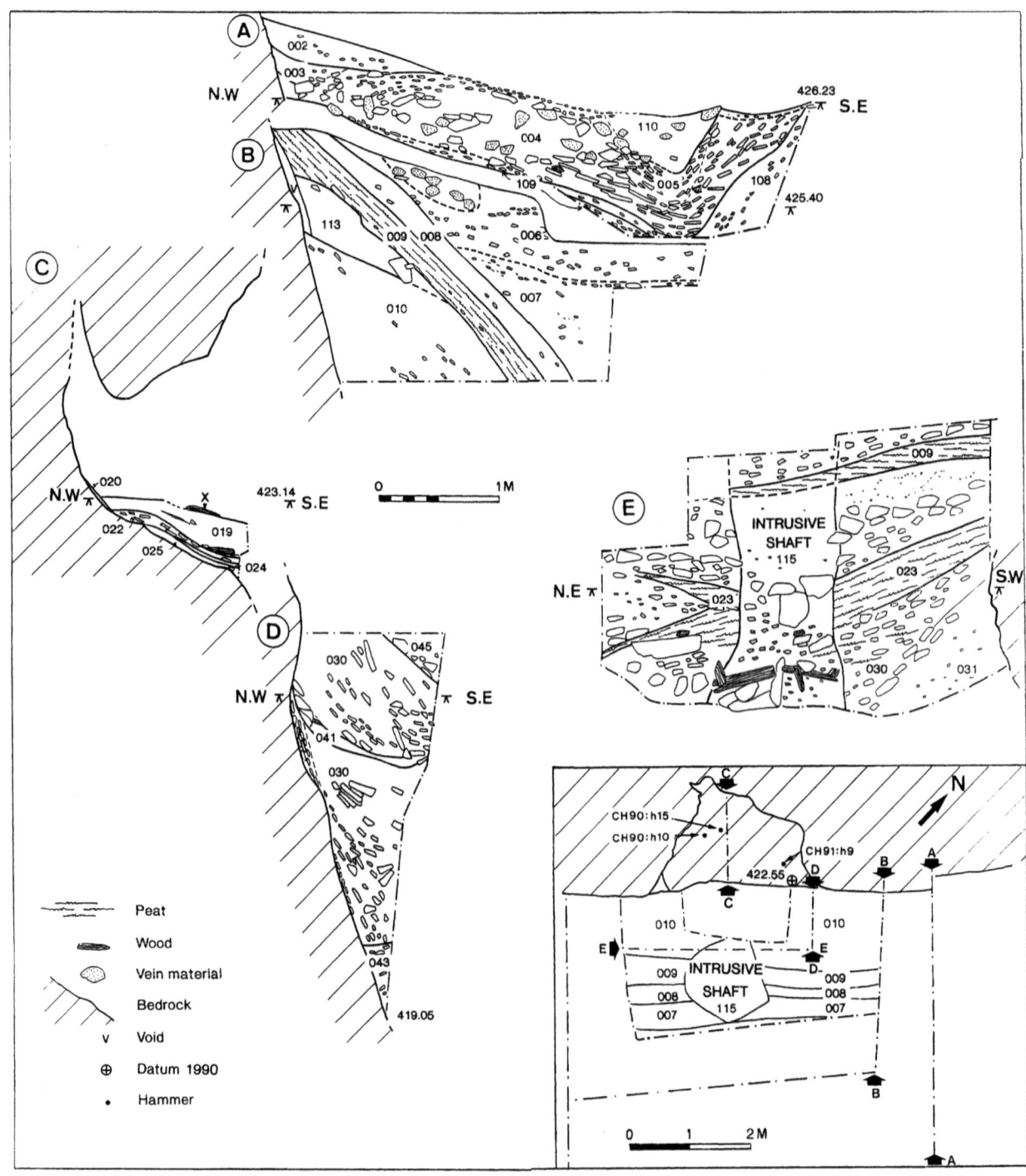

Figure 37: Plan of all excavation cuts, sections and rock outcrop exposed during excavation work carried out within the opencast between 1993 and 1999 (drawing B.Craddock)

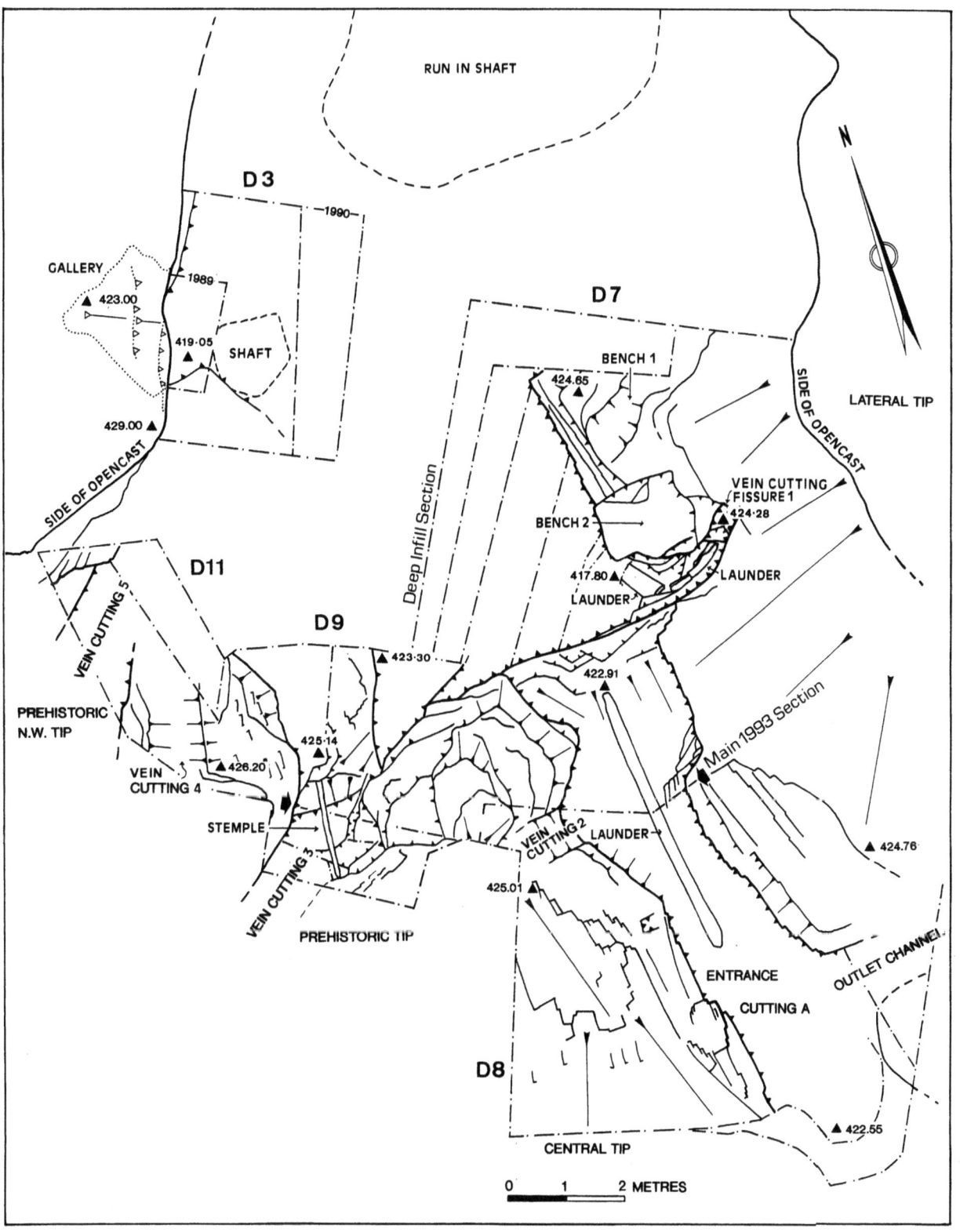

1993 - 1999 - Area excavation of front and south sides of opencast (Mine Entrance, Infill section and Deep Fissure)

Area D7 - Front of Opencast

The purpose of this larger excavation was to section the infilling sediments within the central (deepest) part of the opencast, an area which included possible waterlogged horizons with the potential for the preservation of wooden artefacts. Another objective was the examination of the shallower front of the working where evidence for several rock-cut 'entrances' was already suggested as a result of preliminary diggings in 1989.

An initial area (D7) of excavation (4 m x 7 m) was thus opened up along a SW-NE axis adjacent to the south wall. Just below the surface and some 2 m from the side of the opencast, the remains of a very shallow modern trial shaft and spoil mound (108) were uncovered, yet this feature seems to have been abandoned shortly after it was begun, the 'shaft' penetrating less than 2 m of the archaeological deposit beneath. Machine removal of this continued to a depth of between 2 and 2.5 m within the centre of the trench, revealing the edges of a substantial rock-cut step (Bench 1) projecting some 2 metres out from the side of the opencast, surrounded by peaty sediments. As on the north side of the opencast, these organic sediments appear to originate high up on the side walls of the working as humic layers, which then dip steeply towards the centre of the infill where these develop into much thicker silt and peat sequences. A semblance of the internal landscape of the opencast was revealed during this initial phase of excavation, following removal of the very top of the peat layer (012/013), the latter revealing a further rock-cut step (Bench 2) some 1.5 m lower down, upon which lay a shallow pile of sterile crushed quartz and a large dolerite hammer-stone [CH93:h1]. Less than 2 metres away, and just covered by the peat (at 422.91 m OD), could be seen the very tip of the end of a large timber, subsequently identified as part of the prehistoric mining launder [CH94:w2]. The peat layer surrounding this was recognised as being the same, or else similar, to that identified in 1989 (D2 013/016) overlying Early Bronze Age mine spoil. Following this discovery, excavation over all of this area continued by hand, a stepped section forming the west side of the excavation cut, with rock pillars and the opencast wall effectively forming the other three sides. During August 1993 earth moving machinery was brought in (across the moorland route to the north of the site) in order to help remove the 1-2 metres of surface deposits over a larger area of proposed excavation to be carried out at the front (or lower) end of the opencast. A ramp was cut from the south-west corner, and sites for the tipping of excavation spoil were located some 10 m along the contour to the west, as well as along the crest of the SE Prehistoric Tip.

Examination of the surface of the northern pillar (Bench 1) showed evidence for *in situ* crushing and pounding of the shale surface, with various hammered hollows, some enclosing compressed clay, and several fragments of broken hammer stones, including one struck rock flake.

Similar areas of pounding were evident on the harder quartz-veined southern rock pillar (Platform 3), in one place this appeared as a faint purposeful pattern of three circular hollows just beneath the turf, now partly weathered away. A thin spill of loose modern mine spoil beneath recent scree formed the surface layer of the overlying infill, whilst beneath this various interdigitating lenses of washed shale, sands, shaly silt, and clay could be correlated with some of the highest waterlain horizons examined in 1989 and 1990. Just covering the top of the

Figure 38: SE side of opencast and top of rock pillar (Bench 1) exposed in Area D7 during excavation in 1993. The upper section(s) of the deep infill section can be seen on the LH side plus the top of the as yet unexcavated Deep Fissure 1 (a dark patch) in the bottom RH corner (photo ST)

rock pillars, and dipping steeply into the centre of the opencast, was an organic silt (010), most likely the continuation of the dark soil/silt horizon 009 within the scree sequence embanked against the northern cliff of the opencut, its position perhaps suggesting another high, perhaps a rock ridge beneath the un-excavated area in between. Below further scree, a thin dark horizon (012/013) found clinging to the sides of some of the rock pillars (with faint traces of lichen, grass, moss and roots) developed into a thick sequence of organic peats and silts, in some places more than 2 metres thick (Fig. 41) within the central front area of the opencast. This showed a transition from organic silts (e.g. 012a), to humic woody peats (012b), to silts with sphagnum (013), further leaf and wood peats (013b), and into thick mossy peat sequences (013.1, 013c-e) - the latter section excavated to a depth of more than 5 metres between 1993 and 1996, and serially sampled for pollen, macro-plants, and insects. A radiocarbon date from the top of this infill: 012.a2 [OxA-10042, 1782+/-37: 130 - 330 calAD (Stuiver et al. 1998)] suggested an almost unbroken accumulation of organic sediments from the Middle-Late Bronze Age up to the 2nd. century AD, with additional evidence for charcoal accumulation, ash from grassland burning, and the discard of cut wood waste at a number of different horizons - in particular during the Late Bronze Age and Roman Periods. Much of this activity was probably agricultural, although some small part of it may still have been linked to mining, but apparently not within or near the opencast itself which seems to have been long abandoned by this time. Some of these layers can probably be correlated with other peat horizons found in the mine - for instance 013b with 019 (D3 Mine Gallery), 013c with 016 (D2 sondage), and perhaps between 012b with the peat found covering the tip of the exposed launder.

Within the lower part of this trench (and against the rock wall sides of the pillars) the peat base (013) overlies a mixed layer consisting of scree, shale gravel and re-deposited mine spoil (018); eroded material evidently washed in from the sides of the opencast, and probably representing a slow accumulation over many years following the abandonment of the mine around 1500 BC (calculated on the basis of the lower radiocarbon sequence). Indeed the shale present in the top of this layer, that is just beneath the junction with the peat, appears worn and discoloured, suggesting a well defined but brief phase of erosion and weathering. The layers beneath (058 and 060) contain larger, fresher shale blocks as well as some remnant charcoal, suggesting an input from material naturally breaking off the shattered and fireset rock walls, followed by the slumping-in (or possibly backfilling?) of rock waste,. The rapid accumulation of this deposit has preserved voids within several of the mined-out undercuts, some of which have since become infilled with mineral ochre. From the eastern side of this rock trench a vertical fissure resulting from the removal of a SW-NE branch of the vein narrows from 1.5 metres to about 10 centimetres over a distance of little less than 2 m, whilst the top of this feature (at 423 m OD) appears shaft-like, lying in between a rock bench (Platform 2) and the centre of a what appears to have been the main entrance into the opencast (Entrance A). About 0.5 m from the edge of this the north end of the launder was found.

Figure 39: North end of exposed alder wood launder (047) lying on floor of rock cutting (Entrance A), with top of peat infilled Deep Fissure (1) being excavated in foreground (ST 1993)

Figure 40: Recording the upper section of infilling sediments within the centre-front of the opencast (D7 Deep Infill Section) in 1993. Note several metres of compacted peat at its base. Photo ST.

Figure 41: Deep Infill Section within centre-front of opencast (Area D7). Simplified running section to deepest point reached (in 1999) during excavations carried out inside of the mine. The base of the section is at the forefield of the Deep Fissure 1 vein working (drawing B.Craddock)

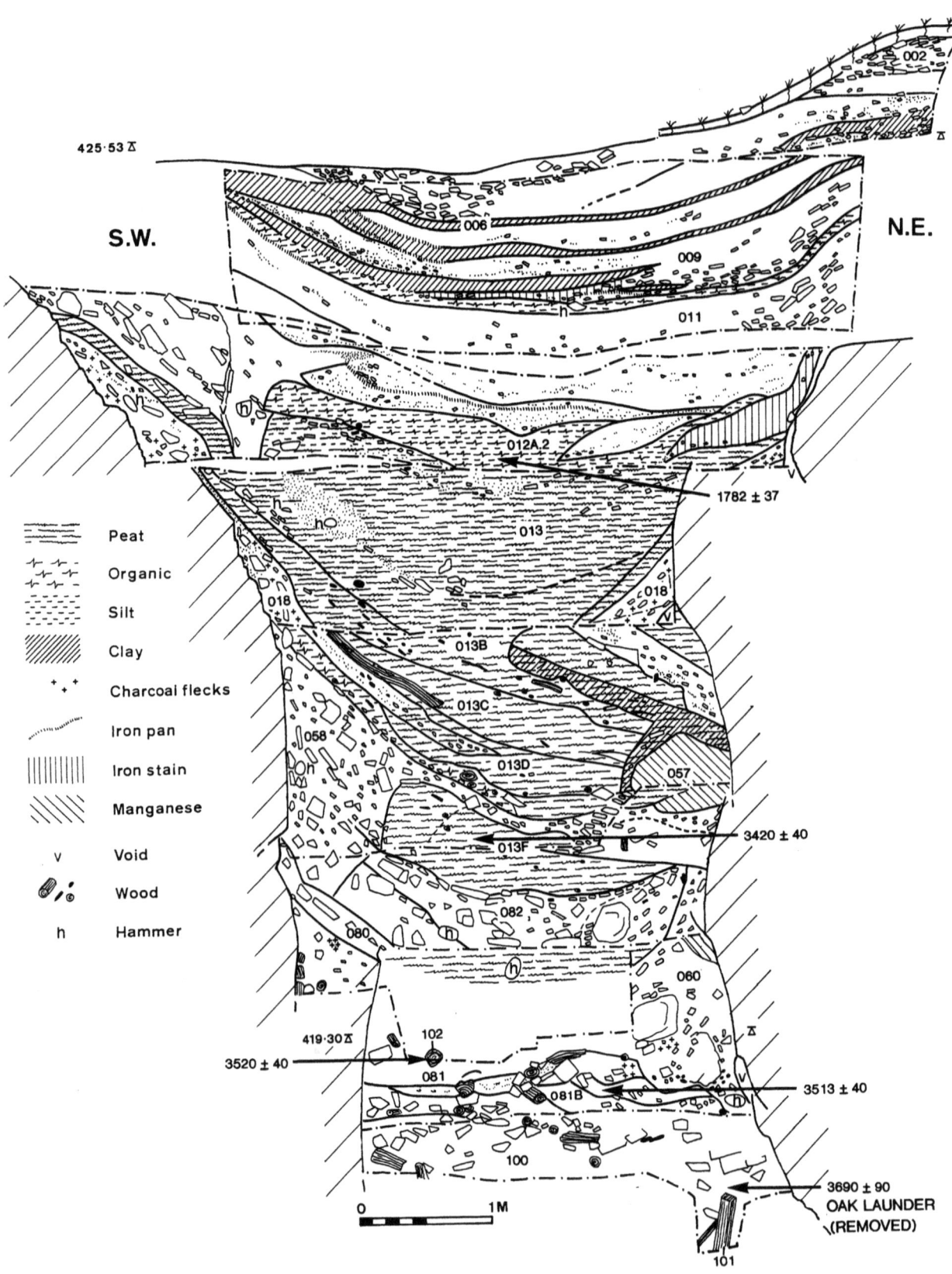

Area D8 - Mine Entrance A

The discovery of this entrance cutting, with what appeared to be a complete wooden drain lying buried beneath the spill of many tons of prehistoric mine spoil, served to shift the focus of excavations from the centre of the working to its south-eastern margin in 1993. Within it there was a strong likelihood of finding *in situ* Bronze Age mining sediments, and perhaps also the waterlogged remains of wooden tools preserved at much shallower depth; thus between 1993 and 1996 this 10 metre long rock channel was excavated in full. From its base the unique 5 metre long alder wood launder was removed in sections, and these are now undergoing conservation at the National Museum of Wales.

One metre of the less well preserved and damaged northern end of the launder was revealed in 1993 (its deteriorated condition at this end probably being due to original exposure and weathering prior to burial with mine spoil following the abandonment of this area of the mine), the cross-section of this being visible within the base of the principle reference section for the deposits infilling the entrance channel (SEE Fig. 45; the position of main 1993 section (D8) can be seen on the plan in Fig. 37). The latter section showed some 3 metres of horizontally layered deposits infilling the stepped sides of a 1.5 -2 metre wide cutting which had been excavated following the cleavage direction of the slate (a method which must have made it easier to remove large amounts of un-mineralised rock). The top of this infill revealed the base of the south end of the 1989 trench D4 at a point where this had cut the edge of the turf-stack dam (031) beneath which lay a buried turf, a thin clay layer, and below that waterlain shale gravels and silts. The upper horizon of this (035) contained some organic material, and thus may well correlate with the top of D7/012 (035), the darker horizon beneath showing soil development (036), and below that pale grey-brown laminated gravels and clays (037-039), the latter with traces of mineral processing residues, and therefore probably equivalent to the base of D4/021 and possibly post- bronze age in date. However, beneath this lay a quite different layer of angular shale scree mixed with some re-deposited mine waste (040) resting upon the distinctive washed-in abandonment horizon (018) of the prehistoric mine, the latter dated to around 1500-1600 BC. The latter seals a sequence of both coarse and fine layers of mine spoil within the base of the channel, some of which contain undisturbed lenses rich in charcoal (041, 044 and 046) therefore almost certainly consisting of *in situ* rather than re-deposited fireset material. The lower layers of blocky mine waste (045), and charcoal-rich crushed vein rock (046) contain considerable amounts of clay infill. The wooden launder appears to lie within, and in places beneath the latter layer, suggesting the dumping here of fresh fireset material following the temporary abandonment of the drainage route and re-commencement of mining within the near proximity. Charring of the upper edges of the wood suggests that hot ashes were raked over the still exposed and recently abandoned launder. Furthermore, its position slightly to one side of the entrance cutting, plus the more compressed nature of the fine grained sediments underneath between it and the rock wall on the western side suggests a pathway worn by the miners, thus confirming that this was probably still in use right up to the time of its final burial. More than one metre of spoil was eventually tipped over this, each charcoal band suggesting three renewed mining episodes. However, sometime following the abandonment of the mine circa. 1500 BC a trench was cut down through these mining layers (following the deposition of 018 but prior to that of the overlying scree (040) and silts) along the length of the entrance, perhaps to re-activate the drainage channel in order to help lower the water level within the dammed and probably flooded opencast, or possibly even to retrieve the still useable launder. This latter trench appeared to miss the launder, and over its detectable length also failed to reached the rock floor of the channel. Nevertheless, it may have reached the level of the rock lip at its northern end, hence free-draining the opencast to its then practical limit. Soon afterwards the trench was fairly rapidly backfilled with loose blocky mine rubble (043), ending what may have been the last prehistoric prospection or mining activity on site.

Figure 42: Excavated Deep Fissure 1 (foreground) with remaining 2 m+ section of alder launder resting on top of basal mining sediments within rock-cut channel Entrance A. View south (photo ST 1995)

It seems plausible that the north end of the launder was designed to rest upon the raised rock lip which forms the southern edge of the Fissure 1 opencut (the presence here of a low rock barrier would clearly have prevented water from re-entering the workings). From this spot the launder must have carried the drainage water (emptied into it by hand or else channelled from another launder section) over an uneven floor to reach an exit point from whence it could be discharged downhill, thereby avoiding any unnecessary flooding to the work area. The use of branch and brushwood supports and flat stones placed beneath this seems to have ensured a gentle gradient of some 2-3 degrees, sufficient for water to flow, the most obvious of these supports being a number of cut branches of birch, oak and hazel (050 (A-D)) placed within the base of a 1 metre diameter rock-cut hollow formed from of the partial excavation of a narrow galena vein.

Figure 43: Brushwood infill + support for base of launder (ST 1994)

The base of this hollow was filled with some 30 cm of fine grained mining sediments consisting of broken shale, clay, spalls from hammer-stones, hazel *(Corylus sp.)* twigs, moss and grass (050), the latter overlying layers rich in finely crushed and discarded galena (052) and beneath that charcoal, quartz and clay (053) associated with areas of pounding on the rock floor (possibly mortar holes used for crushing ore) - all of these pre-dating the final use of the launder above. A fragment detached from the northern tip of this launder gave the following radiocarbon date: BM 2908, 3690±90: 2365-1875 cal BC (or rounded out to 2370-1870 cal BC).

During 1995 and 1996 the remaining 3.5 metres of launder (047 C&D) were excavated and removed along with those mining sediments overlying and underlying it along with the remaining debris infilling the southern end of the entrance channel. Excavation of this whole feature in five 1.5 metre long sectors (AA-AE), divided by three 0.5 metre baulks, allowed better correlation between the thin discontinuous horizons of mining sediments.

A continuation of the same galena vein on the west side of the hollow below the base of a prominent vein-cut (Fissure 2) uncovered a small working area containing crushed copper and lead ore (054), the charred remains of oak *(Quercus sp)* fireset timber, plus several discarded mining tools including a burnt antler pick (062) and the two halves of a broken withy handle for a hammer-stone (076). To the south of the main section (Section 1) it appeared that the base of the launder lay within a shallow trench cut through a bank of fine mining sediment (052), whilst the cut or re-positioning of this drain also pre-dates the main phase of deposition of brushwood (050-051), the latter perhaps strewn as conditions became wetter underfoot, or else in advance of a seasonal campaign of firesetting. Evidence for a further re-cut of the sides of this ditch also supports the notion of re-use of the launder, both here and elsewhere in the mine. Moreover careful positioning of this was also evident within sector AC in the form of trimmed straight sections of branch-wood placed either side of the rounded base as supports. The finely tooled and well preserved southern end of the launder was located in the baulk immediately to the south of this (Fig. 46 A/D), the cut for it finishing after a short distance, but with fragmentary traces of what may have been a second launder continuing for at least a metre beyond.

Figure 44: Detail of remaining section of alder launder lying *in situ* within rock-cut channel Entrance A. Looking south. (photo ST 1995)

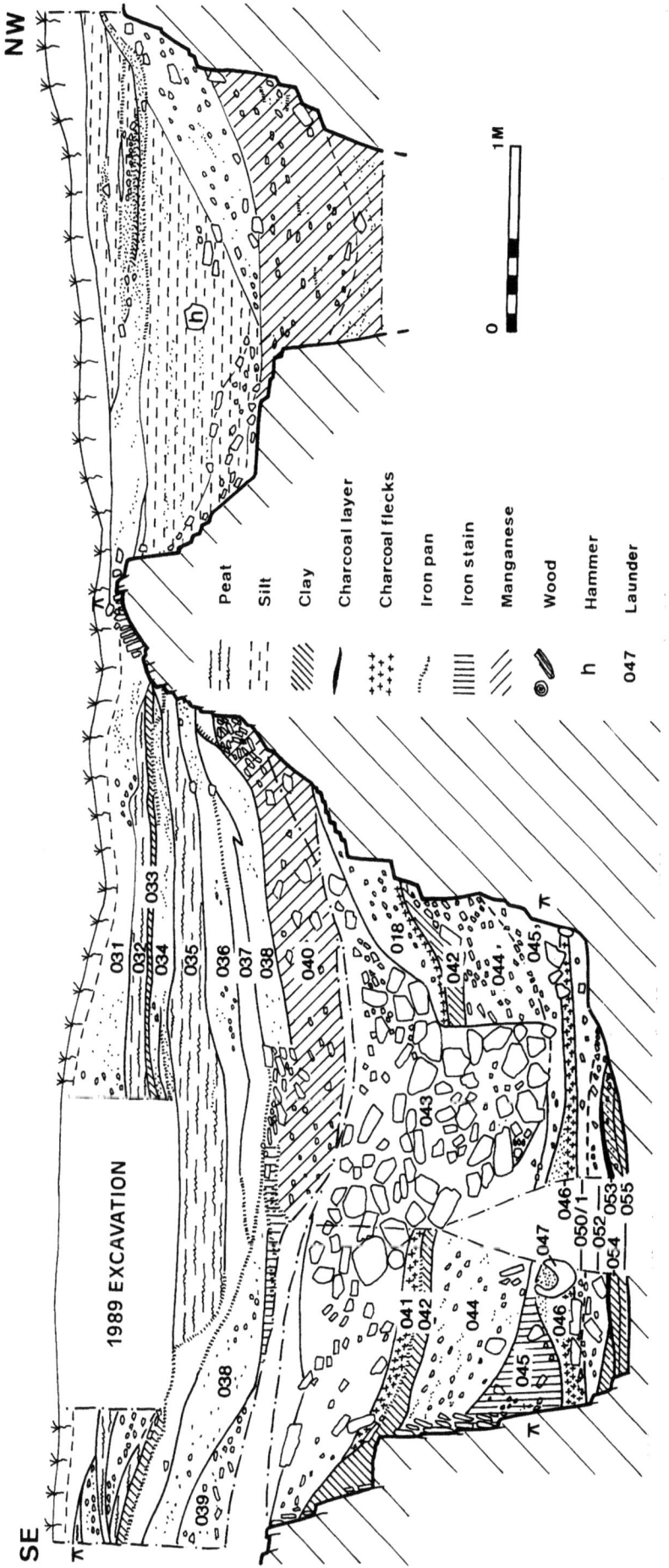

Figure 45: Main cross-section through infill of Entrance A cutting (LH side) and Vein Fissure 3 (RH side) in 1993. South end of D7 excavated area. Shows launder (047) at base of Entrance A. Drawing by B.Craddock.

Figure 46: Plan of Entrance A cutting showing position of wood, antler and hammer-stone finds and cross-sections (between excavated baulks A/B, A/C, A/D and A/E). Drawing B.Craddock.

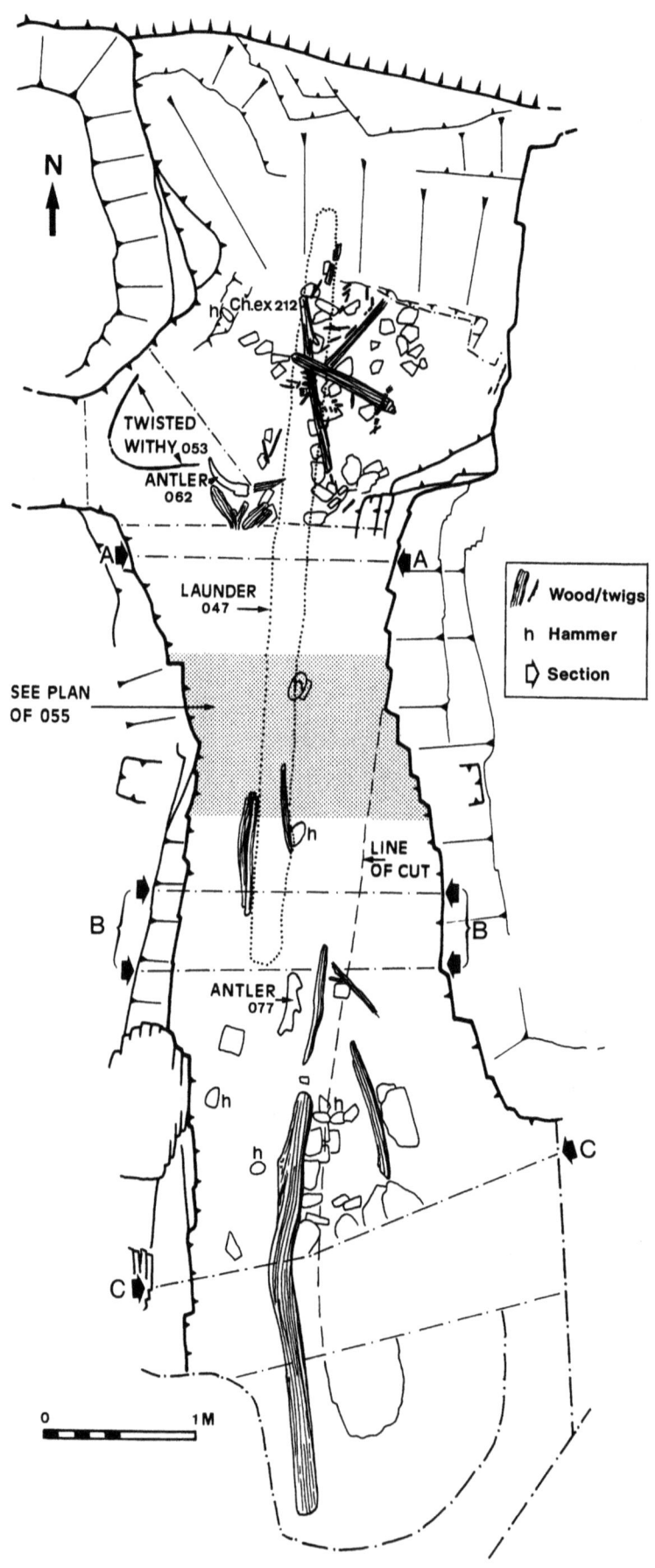

Figure 47: Sections from north to south across Entrance A. Record of baulks left between excavation sectors A/B, A/C, A/D & A/E. Sections are cut through infill (mining layers and overlying spoil) and show position of launder (047) and other timber. Drawing B.Craddock 1995-1996.

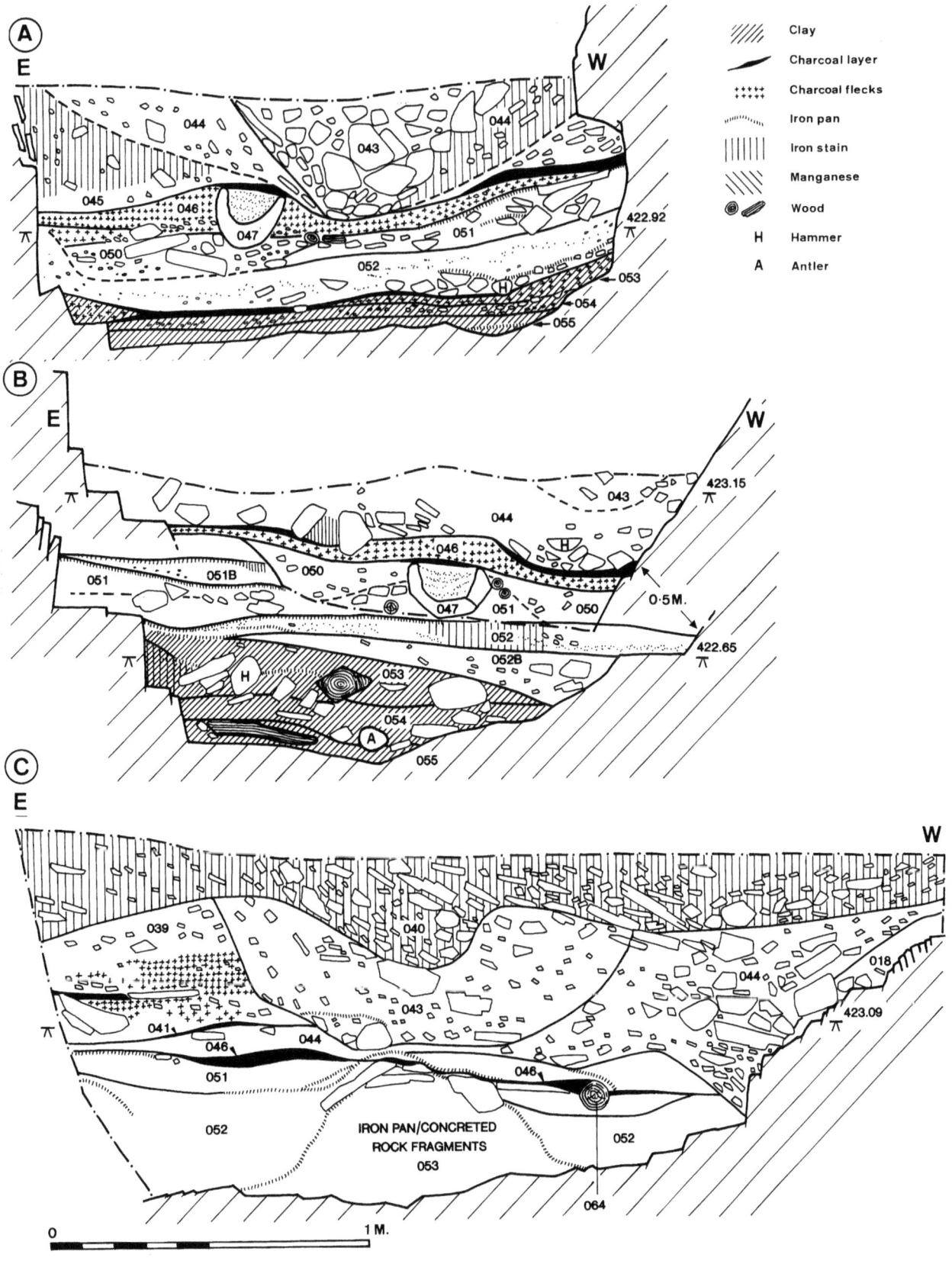

Figure 48: Plan of layer 055 in iron pan showing position of worked wood and withy discarded amongst mine debris. Location of this area (in sector A/C) is shown in Fig. 46. Drawing by B.Craddock 1996.

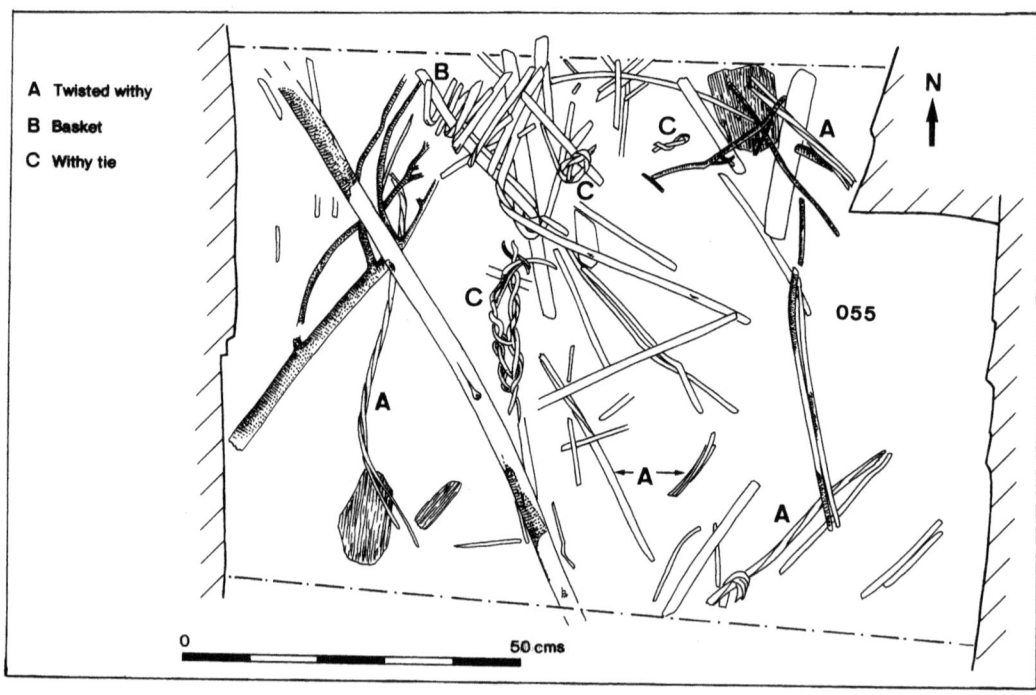

Figure 49: Plan of timber debris deposited within Deep Fissure 1. Spot heights indicated over c. 6.5 m vertical interval. Drawn from plans compiled between 1993-1999, B.Craddock

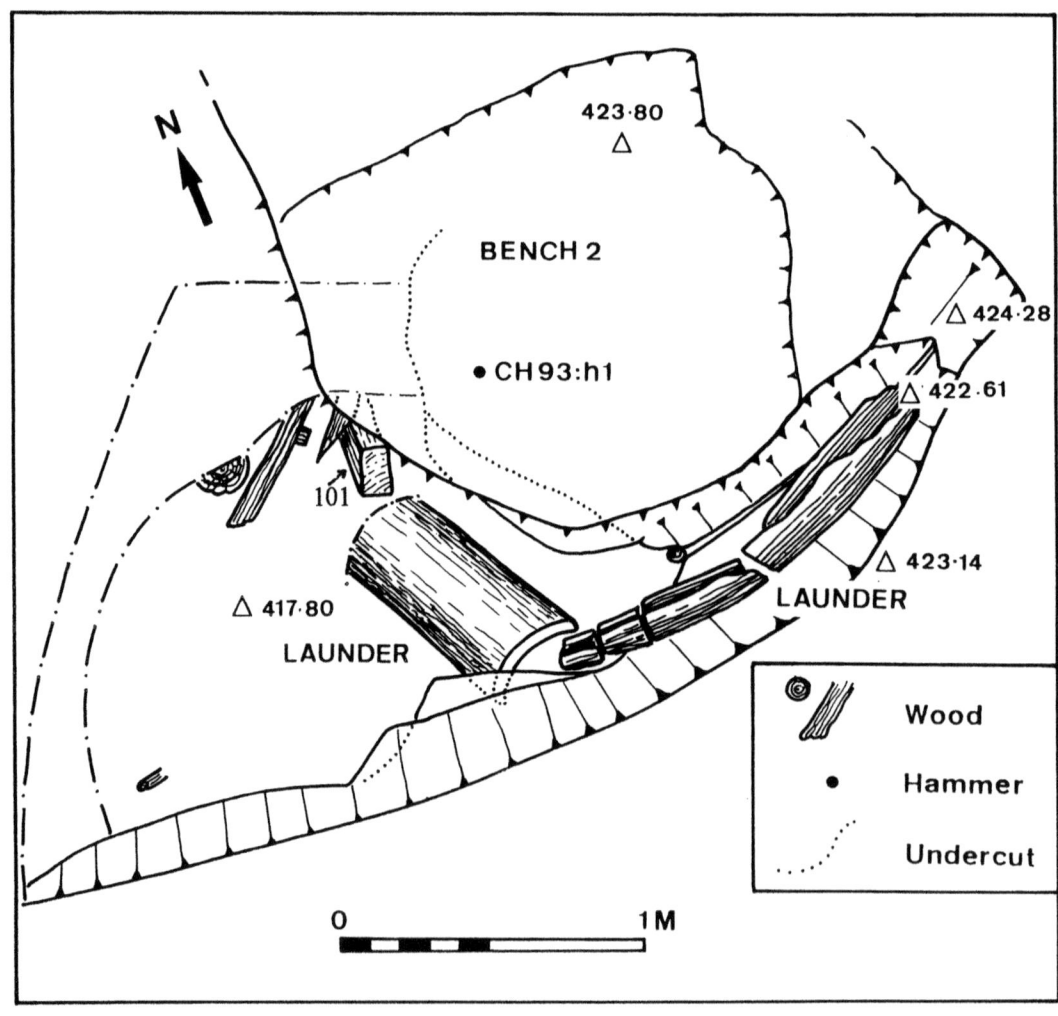

The route which this (putative) second launder took followed the base of the west side of the entrance, its course marked by a shallower cut traceable for another 4 metres, on the east side of which was uncovered a substantial (2.5 m long) oak timber (064), perhaps one of its original supports. The entrance was not excavated beyond this point, although the floor could still be seen sloping steeply away, in all probability ending in the channel now lying buried between the Central and South East Prehistoric Tips, whose existence was proven by excavations in 1986 (SEE Fig.20; D1a). Interestingly enough the gradient of the rock-cut floor of the entrance channel suggests that the original pre-launder drainage exit was in fact through a gap to the east, a position which now closely corresponds to the final resting place (of the southern end) of the remaining launder.

Figure 50: Excavations in progress at south end (exit) of rock-cut channel (Entrance A). Log is in position of removed (?) launder. The ladder marks probable position of original drainage exit (ST 1995)

Excavation during 1996 of the residual layers of mine sediment *below* the level of the launder yielded some of the most interesting small finds. The reason for the good preservation of wood here may come down to the presence of several fine grained compact sediments rich in organic material lying amongst water-filled hollows in the rock floor. Indeed, it seems possible that some of these objects may well have remained in a waterlogged state since they were discarded and trodden underfoot, although problems both of detection and also of excavation accompanied their subsequent removal from what was now an ochreous and in places a well concreted iron pan.

Within a depth of less than 20-30 cm of sediments a certain cyclic repetition of layering, ore fragment types, and artefactual evidence was observed, suggesting that some sort of ore preparation and separation activity was taking place within the entrance. Several layers (056 & 056) overlying the rock-pounded floor, as well as another (052) lying immediately beneath the launder (associated with the extraction of the lead vein in Fissure 2 (SEE Fig. 99)), appeared rich in crushed and oxidised galena fragments. Alongside were larger charcoal fragments plus spalls and flakes resulting from hammer-stone usage. In between these layers, several finer grained yellow-olive green coloured silty sediments had copper staining associated with them, and contained numerous cut and charred fragments of oak firewood, much hazel brushwood, plus a horizon (055) rich in broken, burnt and discarded fragments of twisted withy hammer handles, rope, ties, woven basketry fragments and a fire-stick, all of which were excavated from out of a clay-rich iron pan horizon beneath the central part of the launder (SEE Fig. 46 (plan of sector AC).

Figure 51: Fragments of a coarsely woven hazel basket and withy tie in early stages of excavation from iron ochreous layer (055) Photo ST 1996

Immediately above this a dark grey sandy sediment (054), similar to that recorded in sector AA, yielded traces of copper, much comminuted charcoal, occasional fragments of antler, but little or no evidence for burnt wood. A well preserved antler hammer/pick (Fig. 52) was recovered from this layer in 1995, from a spot just beyond the southern tip of the launder, whilst much smaller antler fragments were found within another similar looking horizon (050a) some 5-6 cm above this. The antler pick (077) produced a rather younger date [OxA-6684, 3405±70: 1890-1520 cal BC (Stuiver et al. 1998)] than the launder lying above it, confirming the relatively late date for the last phase of working in this part of the mine, as well as the probable long currency of use or re-use of the latter.

Figure 52: The handle of a well preserved antler pick/hammer (077) protruding from the base of section 4 (A/D) of the basal mining sediments lying on the rock floor of Entrance A (photo ST 1995)

Area D7 - Deep vein working (Fissure 1)

It seems probable that much of the water draining out of the Entrance (A), as well as least part of the mine spoil, layers of crushed mineral plus mining sediments dumped within the base of this channel were ultimately derived from the exploitation of this deeper working (Fissure 1). Furthermore, the juxtaposition of the channel with its edge makes it clear that both were closely related, the latter perhaps an access route by foot (possibly using ladders?), as well as being a potential haulage-way for lifting out ore, waste rock, and water, and also for lowering timber into the mine. On completion of the excavations in 1999 this near vertical fissure had been cleared to a depth of just over 5 metres beneath the floor of the entrance cutting at its northern end (or a total of 8 metres from surface), with no sign yet of a bottom, nor even of any rock ledges upon its otherwise vertical sides - instead there were indications of increased undercutting of the footwall to the north, something which *may* suggest a considerable depth of buried workings beneath. Further excavation here was impractical, given the confined space remaining between the main infill section and the narrow walls of the fissure, but the use of probing rods indicated at least another 2 m of loose mine spoil/sediment below this. Overall, some 10 square metres of the mineralised hanging wall was exposed within the excavation, along with almost 4 metres vertical height of the vein at the narrow forefield, where the vein began to pinch out, and the rock walls converged. A detailed surface elevation of the hanging wall has been drawn (Jenkins & Timberlake 1997;34), the latter showing numerous tooling marks, including those excavated most probably by antler picks or stone chisels along narrow joints within the adhering iron carbonate vein, perhaps an attempt to pick out the last few areas of copper mineralisation, or else to lever off the hard enclosing quartz vein which had formed a barrier to further mining at this point.

Figure 53: Primitive-looking pick or chisel holes within un-worked galena vein found left adhering to south wall of Fissure 1 (photo ST 1995)

The use of much slenderer tool types was in particular noticeable at the forefield, where minerals, possibly copper-bearing ones, had been carefully extracted from bands little more than 2-3 cm thick either side of the remaining galena. In one place the wall-rock on the left-hand (footwall) side of the vein had been partially removed by hammering, enabling extraction of ore to take place through chiselling and leverage of fragments into the hollow created. Elsewhere, the surface of this lead vein (now heavily oxidised) appears to have been attacked with hammer-stones and picks, but typically this has had little effect on removing the quartz slickenside.

Nevertheless, this same vein had been removed en masse, below a point where it thickened within the forefield, some 4 metres from surface. Here it had been undercut, whilst beneath it lay a deposit of crushed galena intimately mixed with charcoal (061), the latter assumed to have been associated with contemporary firesetting. These charcoal fragments provided the following radiocarbon date [OxA-10043, 3595±45: 2130-1770 cal BC (Stuiver et al. 1998)], one which suggests fairly early exploitation, yet probably also exhaustion of these richer, but deeper vein workings within the mine.

Following the exhaustion of this particular vein the sub-vertical opencut (Fissure 1) may have continued in use as a footway into the mine, although clearly this was no longer used as a route to draw water, the sloping end (closure) of this becoming infilled bit by bit with discarded pieces of wood. From top to bottom this vertical timber pile contained a poorly preserved 2 metre length of alder wood launder (024a+b (SEE Fig. 49) - a piece of somewhat somewhat similar form and dimensions to that recovered from Entrance A of the mine), the latter resting upon a further one metre (20 cm diameter) length of alder timber enclosed within a layer of wood debris including numerous fragments of worked alder and oak (059)).

Beneath this was found a shallower and somewhat wider (31 cm) launder made of oak, in total about 1.5 metres long. The very base of this last launder (083) was found buried amongst scree and other wooden debris and appears to have broken off, the lower half of it perhaps lost within the un-excavated shaly mine waste beneath, underlying what now appears to be the deepest artefact (101) recovered from the mine, a discarded square-cut oak stemple chamfered at both ends for wedging against the rock sides of a crevasse (such as this), in a position where it might have been used as a means of access into the workings, or else as a support for a launder or other timbering. A fragment of the oak launder returned a marginally later C14 date [OxA-10023, 3620±40: 2140-1820 cal BC (Stuiver et al.1998)] than the original alder example, confirming that this area of the mine was probably already abandoned, and thus at least part of the drainage system defunct, sometime after 1900 BC.

The sequence and arrangement of discarding wood inside the mine may yet say something about the order in which these objects, such as the launders (at least one of which may have been located at the southern end of the entrance channel) were removed from use, or else the rate of dumping and backfill of the abandoned fissure. The fact that these appear to have been fairly carefully placed may imply that they were deposited there with the option of recovery, in contrast to rubbish, which would probably have been burnt as firesetting fuel or else discarded on the tips outside of the mine.

Figure 54: Broken sections of alder launder found discarded within abandoned Fissure 1 working (photo ST 1994)

The most recent evidence for mine abandonment, as revealed through the excavation of the basal 2-3 metres of mine infill preserved within the main section centre front of the opencast, has similarly proved interesting. A large basal raft of peat (013f) rich in sphagnum moss, leaf and branch-wood layers which sits between layers of shale scree and slumped-in mine waste (058) and rock spoil beneath, the latter derived from the working of a prominent undercut on the right hand side of this fissure, produced a date which probably represents the cessation of activities within the main opencut trench of the mine [OxA-10022, 3420±40, 1880-1610 cal BC (Stuiver et al. 1998)], yet beneath this evidence for natural sedimentation (081a), waterlogging, and the deposition of *in situ* organics (081b) such as layers of moss (081b) appears almost contemporary with the tipping of fresh spoil (081), consisting of shale, clay, charcoal and half-burnt timber. The latter included a round burnt oak stemple (102) [OxA-10024, 3520±40; 1950-1730 cal BC] of almost identical age to a single moss sample [OxA-10027, 3513±40; 1940-1690 cal BC], suggesting that mining continued here within the upper reaches of the mine circa.1800-1830 BC- as the bottoms became abandoned, flooded, and partly backfilled. Amongst the debris associated with these bottom mining layers were found wood shavings and lighting splints associated with the igniting of firesetting hearths, plus several small fragments of hazel withies, and part of shaped wooden handle or rung (SEE Figs. 77-79).

Figure 55: Recording within base of fully excavated Deep Fissure 1 in 1999. View upwards from c. 4.5 m below ground level. Photo ST.

Areas D9 & D11 - opencut trenches at the front of the mine

Towards the end of the 1999 excavations a more comprehensive investigation was undertaken at the front of the opencast - looking at the area which lay in between the Entrance A (Area D8) and the north wall of the working exposed in 1989 and 1990 (D3). This undertaking commenced with the emptying of the infilling spoil from the narrow and incompletely mined minor lead vein (Fissure 2) located within the west side of the entrance cutting (Entrance A). This exposed the hammered rock face beyond, right up to the point where this descended, still following the barren fault, onto a cut rock ledge or step perched above yet another much larger SW-NE opencut (D9), within which a slightly more substantial vein was worked. The southern end of this new trench, here more than 1 metre wide from wall to wall and excavated to depth of at least 2.5 m below surface, was located almost directly beneath the area of the later turf dam and hushing sluice removed in 1989

(the same cutting shown to the right of Entrance A in Fig. 45). The rock infill of this mined-out vein and the cut of the surrounding wall-rock at this point appeared vaguely reminiscent of a vertical shaft, the sudden narrowing of this vein upon its northern edge resulting in a considerable shallowing and narrowing of the cutting northwards over a distance of more than 4 metres, the floor of the latter dropping over 1 metre in height until the vein here began to widen and split, then dropping as a series of rock cut steps into the main trench - the edge of area D7 Fissure 1(SEE Fig .97). A study of the hammered rock floor and sides of this cutting also provided useful information on the techniques of extracting these thin veins, and moreover demonstrated that it had been worked from two different directions, and possibly at different times; the deeper shaft-like cutting being worked overhand, and probably in an up-slope direction (northwards) from outside of the mine, whilst the narrower vein section had evidently been followed southwards on an earlier occasion from the inside of the mine, and had been worked in an underhand fashion along its sole (floor).

Figure 56: Area D9 showing Vein Fissure 3 fully excavated (within area of opencast) at point where workings from both inside and outside (below) the opencast meet. Ystwyth Valley (looking southwards) in background. Photo ST 1999.

Abrasion and wear over this surface, plus the presence of small amounts of 'stamped in' sediments within the hollows of the floor suggested that the cutting had been used, for a while as one of the access routes (a footway) into the deeper section of the mine beyond, then later was purposefully backfilled, at least at its southern end. Here, some 2.1 m below ground surface, at a point where the rock sides started to be undercut, a round oak stemple beam ((103): 1.85 metre long and 7-14 cm diameter) with one prominently cut and chamfered end was recovered *in situ*, jammed between the rock walls, the widest end close to a small rock-cut niche. An area of abrasion over the centre of this beam suggested some function as a support, although there were no clear marks of rope wear.

The layers of mine spoil immediately underneath and above this beam lay banked up against the eastern wall of the cutting, suggesting that these had been tipped in from this side, perhaps from a working immediately above, or else just to the east (in the vicinity of Entrance A), or behind it. Moreover the orientation of these layers was quite different from the stratification of the overlying tips just a few metres away outside of the mine, similarly they were distinct from the spoil horizons rich in soil, charcoal and hammer stones which lay directly above them (004-005); a fact which suggested that this particular working was probably already exhausted and backfilled some time before the final abandonment of the mine. This is also supported by the radiocarbon dating evidence obtained from the exterior of the oak beam [OxA-10025, 3535±38: 1960 - 1740 cal BC (Stuiver et al. 1998)]; the latter date most likely indicating a period of exploitation sometime around 1850 - 1800 BC.

Just over 3 metres to the north-west of this another trench (D11) 4 m x 2 m wide was cut by machine at approximately 90° to D9. This trench confirmed the existence of two further veins worked in prehistory- opencuts which had apparently remained open in places (or else only partially infilled) right up until Early Medieval period. The largest E-W trending vein and corresponding opencut (Vein Fissure 5) lay adjacent to the north wall of the opencast, and this seems to have been the principle carbonate copper-bearing vein of the mine, the same as that investigated in the excavations of 1989-1991 (D3), the position of the latter lying only 2-3 metres to the north. As in Area D3, the trench failed to reach the bottom of the working, although here both walls of the opencut were clearly visible and thus much closer together, converging into a steep-sided fissure undercut upon its northern side, and barely 1.4 metres wide at its base (less than 1.5 metres below surface). Indeed, the scree and sediment layers infilling this were similar, and rested at a similar angle to those recorded within the upper part of D3; the same Early Medieval buried turf/silt layer (009) being easily recognisable within the sequence. Not unsurprisingly, no mine spoil or evidence for hammer-stone tooling impressions survived upon the rather weathered rock faces within this part of the mine, although at the far southern end of this section the narrower Vein Fissure 4, only 0.5 metre wide at the top, appeared to be much less eroded and full of broken mine rock and mineral (106), including traces of rotted chalcopyrite, at a depth of only 20 - 30 cm from surface. This was not excavated. It was divided by a baulk of rock, less than 2-3 metres wide, from the working to the south-east (D9).

The opening up of these trenches at the front of the opencast allowed for correlations to be made between the various different infill horizons - enabling a better understanding of the vein stockwork to be made, and a reconstruction of the sequence of exploitation.

Figure 57: Oak beam stemple (103) found wedged *in situ* between walls of Vein Fissure 3 at deepest part of working. Photo ST 1999.

Figure 58: South section of D9 (Fissure 3) and small location plan of stemple. Drawing B.Craddock.
Figure 59a + b: Sections of trenches E5 and E7 within Lateral Tip (outside opencast). Drawing B.Craddock 1999.

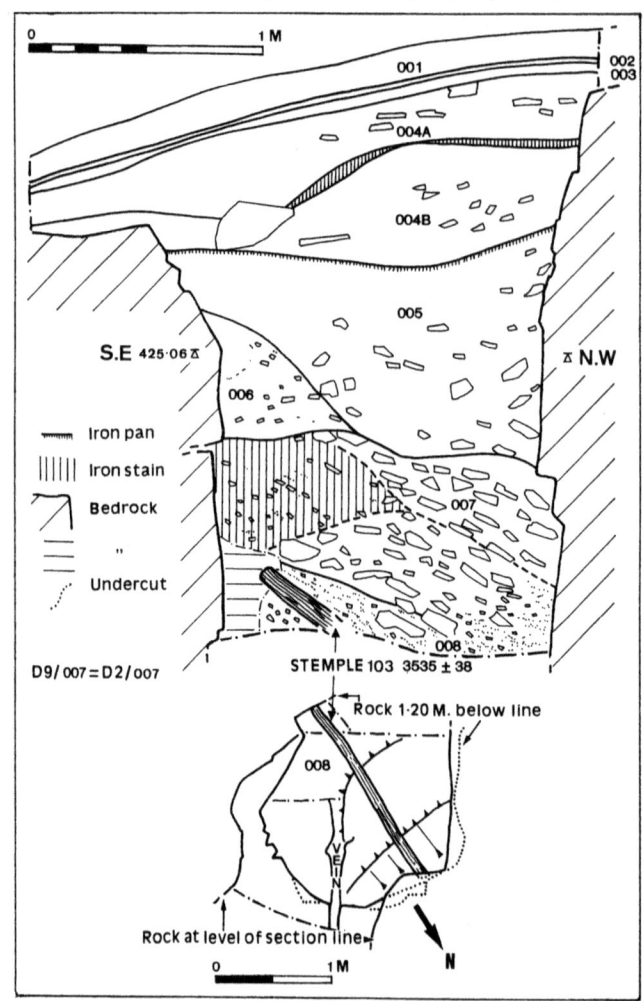

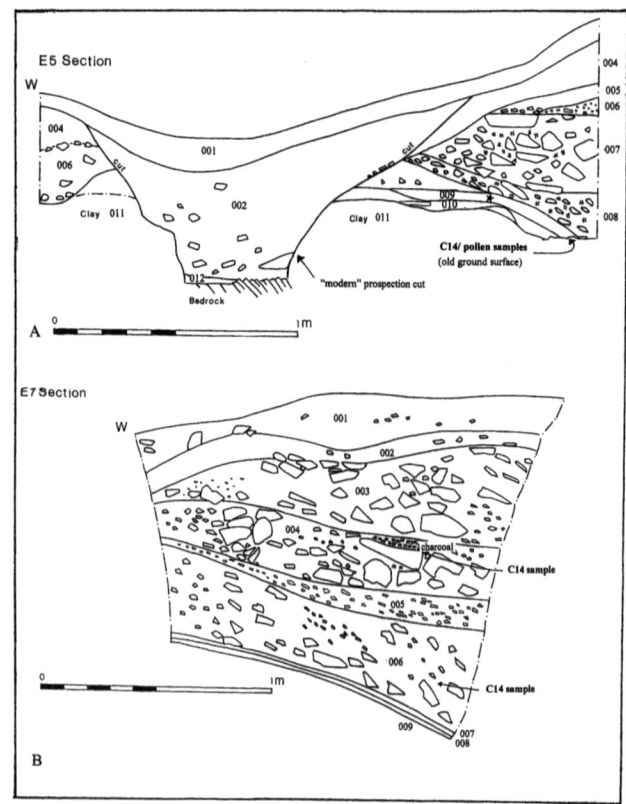

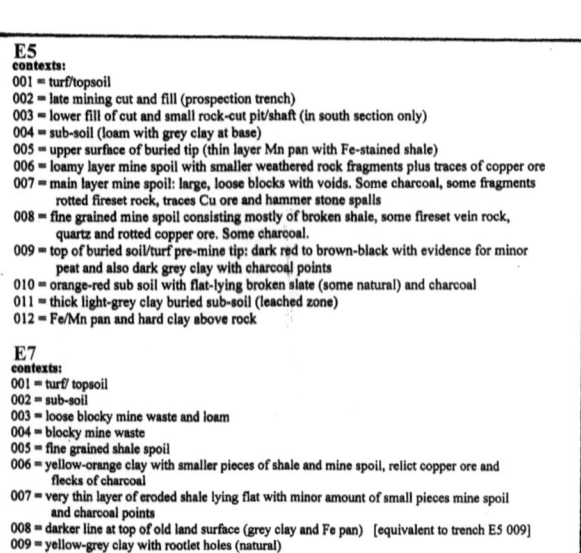

E5 contexts:
001 = turf/topsoil
002 = late mining cut and fill (prospection trench)
003 = lower fill of cut and small rock-cut pit/shaft (in south section only)
004 = sub-soil (loam with grey clay at base)
005 = upper surface of buried tip (thin layer Mn pan with Fe-stained shale)
006 = loamy layer mine spoil with smaller weathered rock fragments plus traces of copper ore
007 = main layer mine spoil: large, loose blocks with voids. Some charcoal, some fragments rotted fireset rock, traces Cu ore and hammer stone spalls
008 = fine grained mine spoil consisting mostly of broken shale, some fireset vein rock, quartz and rotted copper ore. Some charcoal
009 = top of buried soil/turf pre-mine tip: dark red to brown-black with evidence for minor peat and also dark grey clay with charcoal points
010 = orange-red sub soil with flat-lying broken slate (some natural) and charcoal
011 = thick light-grey clay buried sub-soil (leached zone)
012 = Fe/Mn pan and hard clay above rock

E7 contexts:
001 = turf/ topsoil
002 = sub-soil
003 = loose blocky mine waste and loam
004 = blocky mine waste
005 = fine grained shale spoil
006 = yellow-orange clay with smaller pieces of shale and mine spoil, relict copper ore and flecks of charcoal
007 = very thin layer of eroded shale lying flat with minor amount of small pieces mine spoil and charcoal points
008 = darker line at top of old land surface (grey clay and Fe pan) [equivalent to trench E5 009]
009 = yellow-grey clay with rootlet holes (natural)

EXCAVATIONS WITHIN THE LATERAL SPOIL TIP (1994-1999) - THE EARLIEST MINING EVIDENCE:

Shortly after modern investigations started at this site in 1986 it was realised that the ribbon-like spoil mound bordering the lowermost eastern side of the opencast trench must have been associated with one of the earliest phases of activity, for example the tipping of rock waste from workings carried out against the side of the exposed cliff which carried the vein outcrop. Therefore, what is now a fairly large area containing an estimated 400 cubic metres of spoil would almost certainly have ceased to be used for tipping following the excavations of the veins some 1-2 metres below ground level on this side of the opencast, at which time excavations of these trenches downslope, and the resultant tipping of waste downhill would have seemed a more practical and ergonomic alternative.

Investigations in this area began in 1994 with an attempt to re-investigate the northern margin of this tip at its western end, just above the rock lip of the opencast. Here a 3 m x 0.5 m trench (E4) was cut adjacent to the impression of an earlier trench (almost certainly one cut by O.Davies in 1935 and referred to by him as Section B-B (Davies 1947, 59)). The later trench revealed some 85 cm thickness of spoil lying beneath the modern turf line at its southern extremity; consisting of a buried soil overlying shaly mine waste (003) with some evidence of copper staining and carbonate vein rock. Beneath that was a yellow earthy clay (004) enclosing finer mine waste, a little charcoal, and some fragments of rotted chalcopyrite, whilst below lay a horizon of broken shale (005), and finally a dark soil (006) containing isolated fragments of vein-stuff and a lump of galena resting upon the quarried bedrock.

Attention turned to this site again in 1999, with the excavation of 1 metre square sample pit (E6) a few metres to the north of the 1994 trench. Here the overlying turf had eroded away revealing a lens of finely crushed quartz vein-stuff with mineral, apparently cobbed and deposited *in situ* No tools except for a few small flake fragments were recovered, but 3-4 buckets of spoil were passed through a 1 cm sieve in order to collect a bulk sample of the finely crushed fraction. Initial results suggested the presence of many tiny fragments of chalcopyrite/pyrite ore, now barely recognisable following extensive oxidation and leaching. Beneath this, another fine grained loamy layer (002) including cobbed but now rotted chalcopyrite fragments plus minor charcoal, overlay the surface of what appeared to be an earlier tip line, weathered brown on top, with a yellow earthy matrix below, and containing some weathered and well preserved pieces of oak charcoal plus traces of copper ore. The pit was not excavated any further, yet this lower context (003) appeared to be the same layer as E4 004 encountered in 1994. A C14 sample (CH99:12) gave the following date: Wk-9544, 4136 $\pm$58: 2880 - 2570 or 2520-2500 cal BC (Stuiver et al.1998).

Some 6 m up-slope to the north-east the Lateral Tip was sampled near its crest by means of a 2 metre trench. Within this trench the former ground surface, indicated by a grey-yellow clay with root holes, was encountered at a depth of about 0.8 m. Above this lay a thin dark soil line, another of flat-lying washed-out shale fragments, mineral fragments and charcoal, in turn overlain by c. 25 cm of the same weathered yellow clay-rich mine spoil (006). From the latter rotted fragments of carbonate vein rock, copper ore and charcoal were collected, but no hammer-stones. A few small fragments of oak charcoal returned the following date: OxA-10026, 6745$\pm$45: 5730 - 5560 cal BC (Stuiver et al. 1998). Less weathered rocky mine spoil overlay this; in particular a thick layer of blocky void-filled waste with occasional hammer-stones, pieces of charcoal, plus a decayed copper-stained fragment from the end of a broken antler pick. A completely carbonised fragment of oak branch-wood from this layer collected approx. 0.5 metre below ground produced an Early Bronze Age date equivalent to that obtained for the earliest workings within the bottom of the opencast mine [OxA-10044, 3600$\pm$39: 2130 - 1770 cal BC]. However, the much earlier date(s) associated with the base of this spoil tip are not so easy to explain. There seems little doubt though that this earlier spoil was still associated in some way with extraction of copper ore from the vein, if only as a phase of prospection of the exposed mineral outcrop. It is interesting therefore that firesetting also seems to have been employed at this stage. However, the degree of weathering of this spoil suggests that a considerable interval of time must have elapsed before mining recommenced.

At the northern extremity of this tip, opposite the point where the opencast changes direction, another trench (E5) was sunk to sample the depth of mine spoil and date the pre-tip ground surface (above which traces of a buried soil horizon survived). The tip line here was evidently shallower (0.5 m), consisting of alternate layers of blocky, void-filled mine waste, and finer clay and charcoal-rich horizons, but with no evidence for the underlying early phase. In fact a peat sample from the pre-tip ground surface provided a much later date [OxA-10008, 3545$\pm$50: 2030 -1740 cal BC] suggesting that the spoil deposited here may have been generated towards the end of the mining period - perhaps following an eastward extension of the mine.

CHAPTER 7

INVESTIGATIONS OF THE AREA SURROUNDING THE OPENCAST (1993 & 1999)

Earthworks, trenches and the Bronze Age land surface to the south-east of the prehistoric tips (H1-H3, A3-A4 & P1-P3)

In 1993 a series of small pits and trenches (H1-H3) were sunk a few metres to the north and north-west of the location of trench P3 in order to try and identify the former prehistoric land surface and to evaluate its archaeological potential (SEE Fig. 20b). It is believed that these identified a probable buried sub-soil horizon at between 25 and 70 cm below the present land surface, underlying a clay and silt-rich hillwash sediment, and above waterlain shales, solifluction deposits and glacial moraine. No hammer-stone fragments were recovered, even though this area lay within 25 metres of the edge of the spoil mounds. However, traces of mineral fragments and small flecks of charcoal were noted. The buried ground surface here was clearly very uneven, suggesting traces of former digging, though no evidence of former structures were encountered. The edge of one possible stone-filled pit was noted, although this may have been natural.

During 1999 several larger trenches (P1 & P2) were dug to the south of this location in order to sample what may have been an old watercourse, channel, prospection trench or processing area, judging from the presence of what appeared to be crushed shale just beneath the turf. The lower trench (P2) 3.5 m long was cut to intersect a smaller depression, and this identified a thick horizon of buried sub-soil (003) at a depth of between 20-40 cm, towards the base of which were found several small flakes of broken hammer stones plus a minor amount of abraded charcoal. This putative Bronze Age land-surface overlay what appeared to be the natural infill of an earlier or contemporary stream gully. Just to the north of this a much longer (11 m) trench (P1) was dug to try and identify the purpose of one of several linear earthworks. The deepest part of this revealed a similar section consisting of some 20-30 cm of orange sub-soil with broken shale, isolated fragments of burnt rock, several hammer-stone flakes, and flecks of charcoal, overlying 'natural' weathered moraine. Overlying these deposits the earthwork bank consisted of the upcast from the digging of what appeared to be a more recent prospection trench, now peat-filled, but presumably excavated sometime during the late historic period in search of the continuation of the north-south lead/zinc vein up-slope of the small opencast identified at SN 81107512. No dateable artefacts were found within the debris of the latter trench, but another excavation carried out 10 metres to the east of this upon a wider trench and spoil bank (P3) identified the broken stem of an 18th-19th century clay pipe plus several broken hammer stones amongst the upcast - the latter suggesting disturbance of the Bronze Age ground surface or of re-deposited mine debris. The infill of the latter trench was more complex, with evidence for an (as yet) undated hearth within its upper half overlying earlier infill deposits beneath.

Two further trenches (A3 & A4) were dug within the vicinity of the peat drying platform excavated in 1989, a little to the west of P1. It was intended that these would investigate the Bronze Age ground surface beneath the platform and examine its relationship to the buried channel (C). Only the edge of the latter was intersected, yet this did appear to be either contemporary or earlier, infilled with sediment distinctly different from that of the spoil tip, and including in one place, a large lump of broken copper ore. Beneath the peat platform the prehistoric land surface here was shallow and showed no evidence of any pre-existing structures.

To date no clear evidence has been found for processing activity, or any of the associated structures expected of this within the area of gently sloping plateau which lies just to the south of the prehistoric mine and tips. Nevertheless, there are distinct signs here of a rapidly accumulated and disturbed contemporary ground surface, some of which may infer activity beyond the immediate area of the mine.

Geophysical survey (1992) and investigations (1999) of the slope to the west of the SE and Lateral Tips (C5-C9 & D10)

In April 1992 a fluxgate magnetometer survey was carried out by the University of Bradford's Department of Archaeological Sciences of a transect of ground some 40 m x 80 m long to the south of the opencast trench, but including the east and south-eastern edges of the spread of prehistoric spoil. Primarily this was an investigation for areas of burning which might have been associated with processing or metallurgical activity, but was to prove useful also in the location of buried pits and trenches and in defining the edges of the shallow mine spoil cover (*pers. com.* P.Budd & D.Gale). Unfortunately, this failed to reveal the presence of any hearths, and no new pits or trenches, most of the latter already being visible at surface, and now readily identifiable as features of 'recent' prospection. However, the plots did reveal a more extensive sub-surface distribution of spoil, with a much clearer definition between the tipping of the earlier Lateral Tip (with a possible stone kerb delineating its south-east end), the South East Tip, plus another spread of possible re-deposited tip/waste some 20 metres to the east. A long transect between the eastern margin of this tip and an apparently natural rise in between the latter two areas was sampled by trenching in 1999.

The main section of this was formed by a single 9 metre long NW-SE trench (C5) which revealed a progressive shallowing of these spoil layers from > 1.5 m at the west end to less than 40 cm in the east, it being evident from this that the spoil here had filled a major depression within the land surface below the opencast, rather than being deposited above level ground. Charcoal and hammer-stones were recovered from some of these layers now recognised to be highly representative of the sort of waste material left by mining, whilst from a sondage at the eastern end (C5c) a buried turf line was encountered overlying a typical orange sub-soil and a natural deposit of large boulders.

At the eastern extremity of this trench lay the edge of what appeared to be a small pit cut into the natural clay and shale moraine *underlying* this buried turf, and thus pre-dating the mining activity. This pit was also picked up in another trench (C6), the latter cut at a 90° angle some 0.5 m to the east, its purpose being investigate a more recent hollow, probably a surface prospection feature, some 4 metres downslope. The earlier (prehistoric?) pit took the form of an irregular oval-shaped cut some 3 m long, 1 m wide and approx. 60 cm deep within which a large rectangular/lozenge shaped slab of gritstone had been vertically placed with its top projecting no more than 25-40 cm above the (pre-mining) Bronze Age land surface, barely visible from a distance. Excavation of the pit fill provided us with few clues as to its purpose or origin, except to show that this appeared to have been dug to insert the stone, which was then levelled horizontally by means of stone packing around its base, following which the pit was completely backfilled with compacted clay, amongst which traces of rootlets and small burrows were now visible. No finds were recovered apart from a few very small grains of charcoal and two or three weathered fragments of mineral vein. Shattered or crushed fragments of mine spoil were conspicuously absent, providing no indication therefore of contemporary mining or prospection activity.

Immediately downslope this, nestled against the south face of the upright slab inside of a 40 cm deep hollow surrounded by shale equivalent to the mine spoil horizon (004), was found a large ovoid/spherical shaped beach cobble (Chex:459). The latter must have been brought to the site for use as a tool, yet the only evidence for working on it seemed to be a small area of abrasion upon its flattest side, presumably as a result of its brief use as an anvil stone. Fragments of broken stone filling this hollow appear to have come from the 'placed' gritstone slab, and may naturally have spalled-away as a result of frost shattering rather than from hammering. Nevertheless, the possibility remains that both the stamped floor of the hollow as well as the cobble were part of a small crushing bench, or other related feature associated with mining, although just as conceivably this area could have been eroded out by sheltering grazing animals, such as sheep. More certainly, that the buried orthostat (or 'placed stone') pre-dates this, although its function or original purpose remains a mystery. The long axis of this slab (1.2 m wide x 0.8 m deep x 0.3 m thick) faces south-west and overlooks the Ystwyth Valley.

Figure 60: Placed stone slab located in Trench C6 showing excavated pit behind it. View SE (downslope).
Photo ST 1999

Three of the other small trenches/test pits (C7-C9) investigated the context of several other large boulders within the vicinity, but all of these proved to be *in situ* glacial erratics embedded within the underlying shale moraine and solifluction deposit. One further trench (D10) some 25 m north of this point was excavated to sample the eastern extremity of the Lateral Tip beneath the floor of the peat sledding track, and this revealed some 20-30 cm of mine spoil with hammer-stone fragments, but little or no charcoal, overlying an old clay land surface.

It seems unlikely that significant areas of ground still remain within the vicinity of the mine worthy of sampling for the evidence of an associated work camp, yet one should perhaps be cognisant also of the continuation eastwards of the small plateau below the mine which lies between the 404 m - 410 m contours (although this now seems quite disturbed by later surface works), and of the large boggy area on the plateau above (>440 m).

Figure 61: Plan and SW section of Trench C6 with Trench location plan for Site C (1999). Drawing B. Craddock.

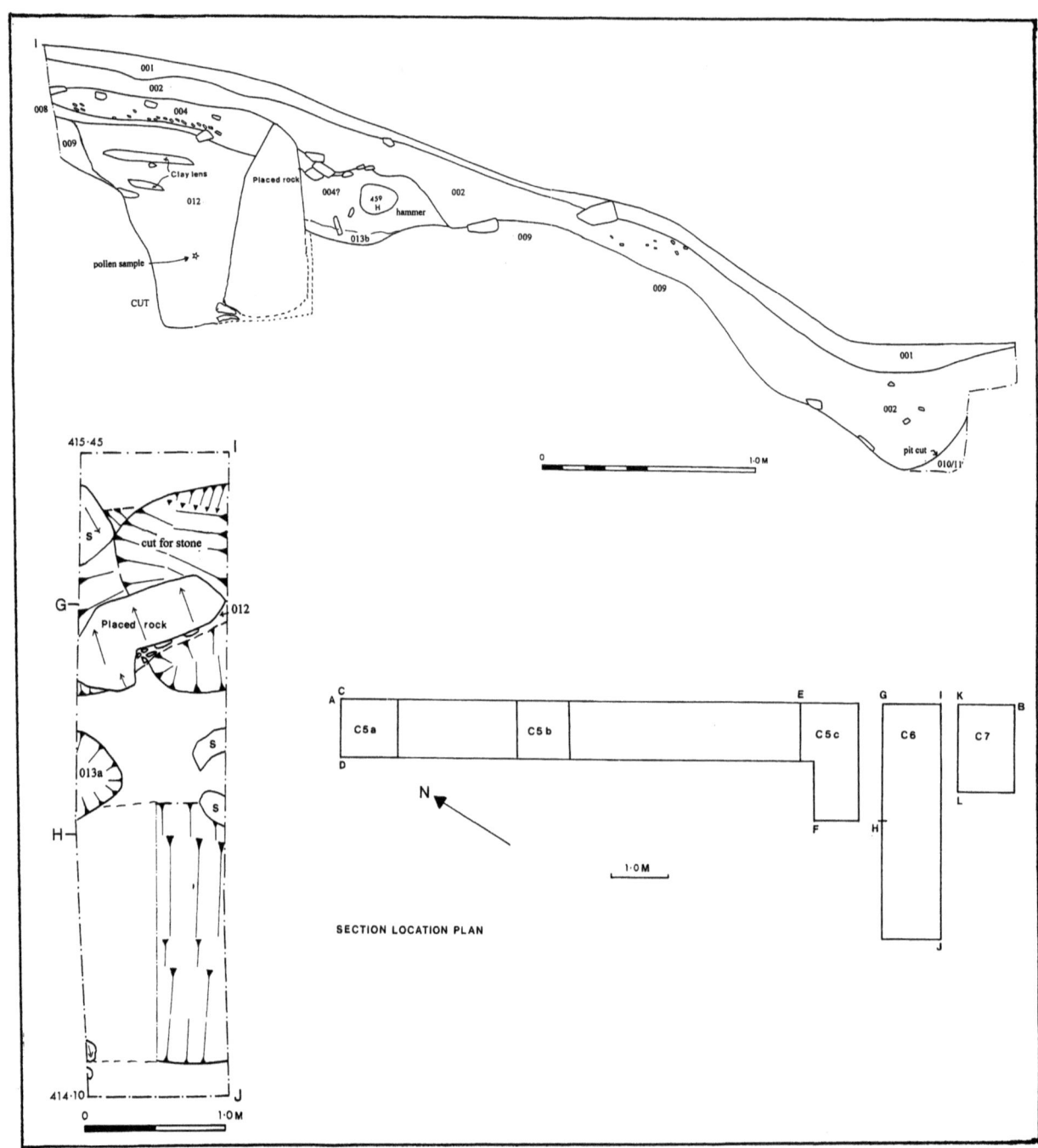

CHAPTER 8

RADIOCARBON CHRONOLOGY (Janet Ambers)

A series of radiocarbon determinations for Copa Hill have been made at a number of radiocarbon laboratories using both conventional (BM, Q, Wk and Beta results) and AMS (OxA results) radiocarbon techniques. These are listed with calibrated age values in Table X, above, and shown graphically in Fig X and X+1. For details of measurement methods used see Ambers and Bowman 1998 for British Museum (BM) results, Hedges *et al* 1989 for Oxford (OxA) results, Switsur 1981 for Cambridge (Q) results, http://www.radiocarbondating.com for Waikato (Wk) results and http://www.radiocarbon.com/labmethods.htm for Beta Analytic (Beta) results.

The vast majority of the determinations were made on wood or wood charcoal as the acidic nature of the soil at the site meant that little other datable material survived. In the past reliance on such plant material at similar sites has lead to doubts being expressed as to whether the dates truly reflect the antiquity of mining activity, rather than the result of the use of fossil wood from the surrounding peat deposits (Briggs 1983). For Copa Hill, however, one determination on collagen (OxA-6684) was possible. This came from a section of a worked red deer antler directly related to the mining activities during the second part of the main mining period, and gave a result in accord with the wood and charcoal results from this period. Unlike wood, antler is not preserved within peat, and this cannot represent the use of fossil material. Similarly the date for a sample from a large worked tree trunk (BM-2908; *Alnus* sp), for which it is extremely unlikely that fossil wood could have been used, agrees well with the wood and wood charcoal results for the first main mining period. These agreements, together with the consistency of the results from both of the main mining periods, provide a convincing argument for a Bronze Age origin for the dated workings. This interpretation is further supported by the dates for the post-abandonment infills which succeed those for the mining activity.

Table 1: Sample contexts, dates and calibrated age ranges for samples analysed (J.Ambers)

	C14 result BP	Context	Material	Possible calibrated calendar age ranges (in calendar years)	
				68% confidence*	95% confidence**
Early mining					
1, Earliest mining					
OxA-10026	6745 ± 45	CH99:10 (E7/006) bulk sample from base of lateral tip	Charcoal (*Quercus* sp)	5715 BC - 5680 BC or 5670 BC - 5620 BC	5730 BC - 5610 BC or 5590 BC - 5560 BC
Wk-9544	4136 ± 58	CH99:12 (E6/003) from base of lateral tip	Charcoal (*Quercus* sp)	2870 BC - 2800 BC or 2780 BC - 2770 BC or 2760 BC - 2620 BC or 2610 BC - 2590 BC	2880 BC - 2570 BC or 2520 BC - 2500 BC
OxA-10044	3600 ± 39	CH99:9 (E7/004) from lateral tip; fragment branchwood	Charcoal (*Quercus* sp)	2020 BC - 1990 BC or 1980 BC - 1880 BC	2130 BC - 2080 BC or 2040 BC - 1870 BC or 1850 BC - 1810 BC or 1800 BC - 1770 BC

2, Main mining period (1)					
BM-2908	3690 ± 90	D7, hollowed out tree truck presumably used as launder	Wood (*Alnus* sp; probably *Alnus glutinosa*)	2200 BC - 1940 BC	2400 BC - 1750 BC
OxA-10023	3620 ± 40	CH99:W1 (D7/081/100/083) hollow tree trunk launder	Wood (*Quercus* sp)	2040 BC - 1910 BC	2140 BC - 2070 BC or 2050 BC - 1880 BC or 1840 BC - 1820 BC
OxA-10043	3595 ± 45	CH95:31 (D7/061) *in situ* debris from worked lead vein	Charcoal	2020 BC - 1990 BC or 1980 BC - 1880 BC	2130 BC - 2080 BC or 2050 BC - 1870 BC or 1850 BC - 1770 BC
3, Main mining period (2)					
BM-2732	3500 ± 50	D4/27, early tip material at front of mine	Charcoal (immature *Quercus* sp)	1890 BC - 1740 BC	1950 BC - 1680 BC
BM-2812	3460 ± 50	D3, fireset debris on ledge in opencast	Charcoal (immature *Quercus* sp)	1880 BC - 1840 BC or 1830 BC - 1790 BC or 1780 BC - 1730 BC or 1720 BC - 1690 BC	1920 BC - 1630 BC
OxA-6684	3405 ± 70	entrance cutting D8	Collagen from red deer antler hammer/pick	1870 BC - 1840 BC or 1810 BC - 1800 BC or 1780 BC - 1600 BC	1880 BC - 1520 BC
OxA-10008	3545 ± 50	CH99:33 (E5.009), pre-mining at top of Lateral Tip	Soil	1950 BC - 1860 BC or 1850 BC - 1770 BC	2030 BC - 1990 BC or 1980 BC - 1740 BC
OxA-10024	3520 ± 40	CH99:W3 (D7/081/102) charred mine wood	Wood (*Quercus* sp)	1890 BC - 1750 BC	1950 BC - 1730 BC
OxA-10025	3535 ± 38	CH99:W4 (D9/008-007) 103 *in situ* wooden stemple	Charcoal (*Quercus* sp)	1920 BC - 1860 BC or 1850 BC - 1770 BC	1960 BC - 1740 BC
Q-3076	3220 ± 70	D1/3, spoil tip outside mine	Charcoal	1600 BC - 1560 BC or 1530 BC - 1410 BC	1690 BC - 1370 BC or 1340 BC - 1310 BC
Q-3077	2990 ± 190	D4, spoil tip outside mine	Charcoal	1430 BC - 970 BC or 960 BC - 940 BC	1700 BC - 800 BC
Q-3078	3210 ± 50	D1/2, spoil tip outside mine	Charcoal	1520 BC - 1425 BC	1610 BC - 1390 BC

Natural Infills					
1, During mining					
OxA-10022	3420 ± 40	CH96:28 (D7/013f) peat layer	wood (*Betula* sp) buried in peat	1860 BC - 1840 BC or 1770 BC - 1680 BC or 1670 BC - 1660 BC or 1650 BC - 1630 BC	1880 BC - 1610 BC
OxA-10027	3513 ± 40	CH99:21 (D7/DF/081) mine debris in base of flooded mine	moss	1890 BC - 1740 BC	1940 BC - 1730 BC or 1710 BC - 1690 BC
2, Post abandonment					
BM-2733	3070 ± 50	D2/13, natural infill of abandoned opencast (front)	Peat	1410 BC - 1290 BC or 1280 BC - 1260 BC	1440 BC - 1210 BC or 1200 BC - 1190 BC or 1180 BC - 1160 BC or 1140 BC - 1130 BC
BM-2759	2850 ± 80	D3/019, floor of abandoned mine gallery (infill)	Leaf and moss peat	1190 BC - 1180 BC or 1130 BC - 900 BC	1260 BC - 1230 BC or 1220 BC - 830 BC
BM-2780	950 ± 50	D3/009 buried turf layer within fill of opencast	Plant material (incl oak leaves)	1020 AD - 1070 AD or 1080 AD - 1160 AD	1000 AD - 1220 AD
OxA-10042	1782 ± 37	CH94:1 (D7/012.a2) organic silt	Fine charcoal extrated from silt	130 AD - 150 AD or 170 AD - 200 AD or 210 AD - 340 AD	130 AD - 350 AD
Later features					
BM-2760	830 ± 140	D3 base of backfilled shaft	Wood (immature *Quercus* sp)	1030 AD - 1290 AD	900 AD - 1450 AD
BM-2828	40 ± 30	D3/027 collapsed shaft shoring within base of prospection shaft	Saw marked timber	1700 AD - 1720 AD or 1810 AD - 1830 AD or 1880 AD - 1920 AD or 1940AD - modern	1690 AD - 1730AD or 1810 AD - 1920AD or 1940 AD - modern
Wk-9543	387 ± 39	CH 93:21 (D7/032) base of turf-stack hushing dam	peat	1440 AD - 1520 AD or 1590 AD - 1620 AD	1430 AD - 1530AD or 1550 AD - 1640AD

Mining related features outside the opencast					
Beta-140992	820 ± 60	PSS1:005, Penguelan lead smelting site	Charcoal (*Quercus* sp)	1160 AD - 1280 AD	1030 AD - 1100AD or 1110 AD - 1300AD
OxA-10041	789 ± 33	CH93:18 (F5/004), leat infill	Peat	1220 AD - 1275 AD	1190 AD - 1290AD
OxA-10045	704 ± 34	CH99:77, Platform Site 1	Charcoal (*Quercus* sp)	1270 AD - 1310 AD or 1370 AD - 1390 AD	1250 AD - 1330AD or 1350 AD - 1390AD
Peat columns					
GrN-17635	2395 ± 35	site CH2 104-105 cm depth	Peat	520 BC - 390 BC	760 BC - 680 BC or 660 BC - 640 BC or 550 BC - 390 BC
GrN-17636	3470 ± 35	site CH2 133-134 cm depth	Peat	1880 BC - 1840 BC or 1830 BC - 1790 BC or 1780 BC - 1730 BC	1890 BC - 1680 BC
Beta-154444	4240 ± 70	site CH3 178-180 cm depth	Peat	2920 BC - 2850 BC or 2820 BC - 2680 BC	3020 BC - 2580 BC

All results calibrated using INTCAL98 (*Radiocarbon*, 40(3), 1041-1084) and OxCal v3.5 (Bronk Ramsey, 1995; for this program version see http://www.rlaha.ox.ac.uk/orau/06_01.htm). * equivalent to ± 1σ ** equivalent to ± 2σ

NOTE (ST): The calibrations shown above are slightly revised from those quoted elsewhere within the body of this excavation report. The latter have been retained and referenced since they have been quoted in other published accounts.

Table 2: calibrated date ranges for Copa Hill shown as probability distributions (J.Ambers)

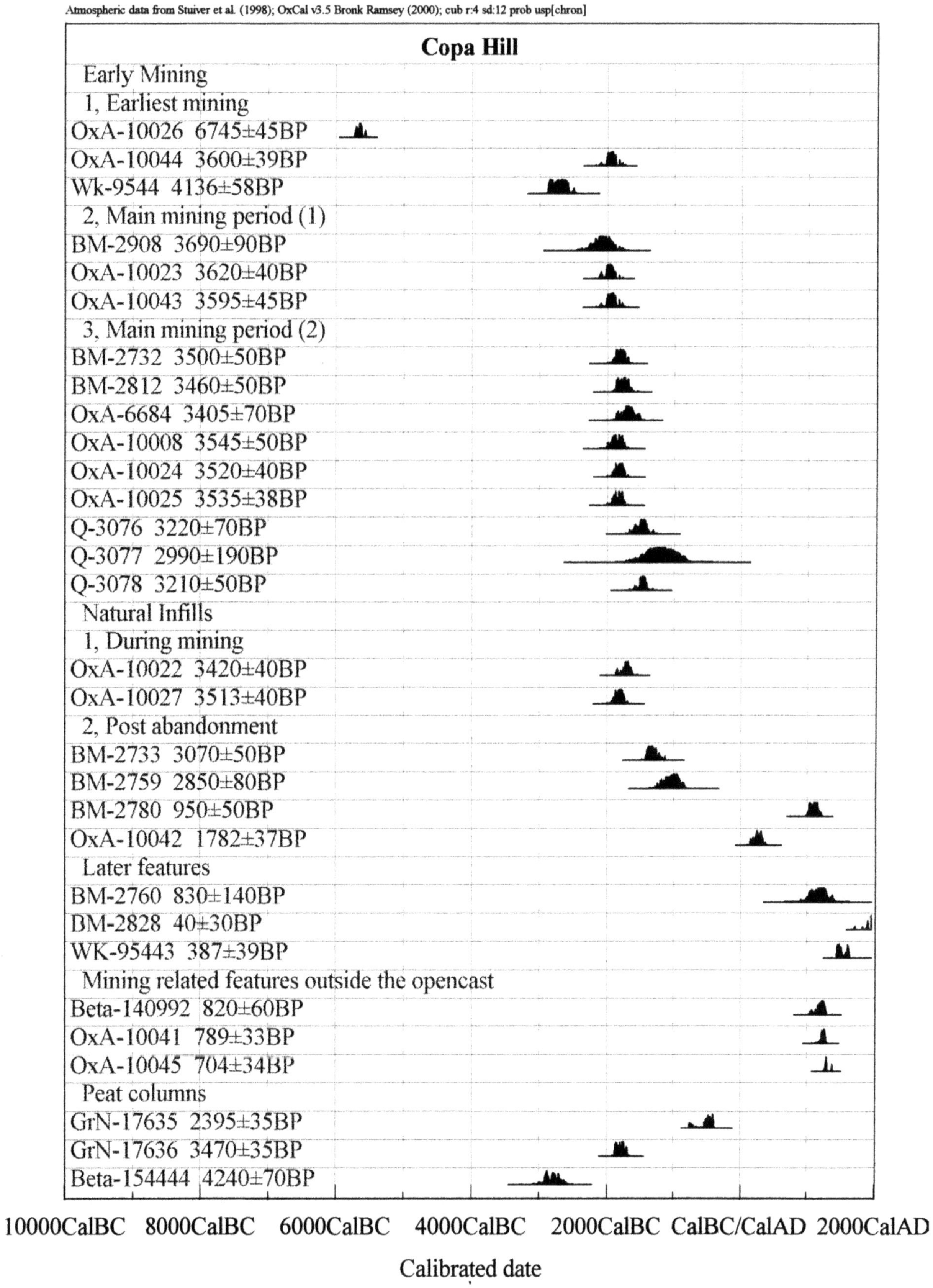

Atmospheric data from Stuiver et al. (1998); OxCal v3.5 Bronk Ramsey (2000); cub r:4 sd:12 prob usp[chron]

Copa Hill

Early Mining

1, Earliest mining
- OxA-10026 6745±45BP
- OxA-10044 3600±39BP
- Wk-9544 4136±58BP

2, Main mining period (1)
- BM-2908 3690±90BP
- OxA-10023 3620±40BP
- OxA-10043 3595±45BP

3, Main mining period (2)
- BM-2732 3500±50BP
- BM-2812 3460±50BP
- OxA-6684 3405±70BP
- OxA-10008 3545±50BP
- OxA-10024 3520±40BP
- OxA-10025 3535±38BP
- Q-3076 3220±70BP
- Q-3077 2990±190BP
- Q-3078 3210±50BP

Natural Infills

1, During mining
- OxA-10022 3420±40BP
- OxA-10027 3513±40BP

2, Post abandonment
- BM-2733 3070±50BP
- BM-2759 2850±80BP
- BM-2780 950±50BP
- OxA-10042 1782±37BP

Later features
- BM-2760 830±140BP
- BM-2828 40±30BP
- WK-95443 387±39BP

Mining related features outside the opencast
- Beta-140992 820±60BP
- OxA-10041 789±33BP
- OxA-10045 704±34BP

Peat columns
- GrN-17635 2395±35BP
- GrN-17636 3470±35BP
- Beta-154444 4240±70BP

10000CalBC 8000CalBC 6000CalBC 4000CalBC 2000CalBC CalBC/CalAD 2000CalAD

Calibrated date

CHAPTER 9

THE PALAEO-ENVIRONMENTAL EVIDENCE (PLANT REMAINS, BEETLES AND WOOD)

PLANT MACRO-FOSSILS (Astrid Caseldine, University of Wales, Lampeter))

Some 23 individual or bulk samples containing leaves, stems and seeds were sampled in 1990 and 1995 from organic horizons within the mine. The samples were allowed to soak then sieved and the fractions > 1mm, > 500 um, > 250 um, and > 125 um collected. Identifiable plant remains were extracted from the coarser fractions and the finer fractions were scanned. Identification was by comparison with modern reference
material and standard texts. Nomenclature follows Stace (1991). Plant macrofossils are generally good indicators of local conditions being less easily transported over long distances than pollen.

Front of opencast
D2/013: Monocot. and dicot. leaf fragments, *Sphagnum* moss, *Carex spp.* (sedges), *Montia fontana* (blinks), *Juncus sp.* (rushes).
D2/016 : Monocot. and dicot. remains, abundant Poaceae (grasses), moss.

N. wall opencast
D3/009 (Early Medieval) :Monocot. remains, Poaceae (grass) seeds, *Calluna vulgaris* (heather) shoots, flowers and seeds, *Juncus sp.*

Mine gallery infill
D3/019 upper A (Late Bronze Age): moss, *Calluna vulgaris* shoots and flowers, *Betula sp.* (birch), *Vaccinium vitis-idaea* (cowberry).
D3/019 leaf horizon: leaves of *Salix sp* (willow), *Betula sp* (birch), *Quercus sp.*(oak), *Corylus* (hazel) or *Alnus sp* (alder), moss (not *Sphagnum*)
D3/019 lower B: moss, dicot. leaf fragments, *Juncus sp.*
D3/024: moss, *Calluna vulgaris* flowers,shoots and seeds etc., *Betula sp* fruit and cone scales, Lamiaceae (thyme) carbonised fragments and *Juncus sp.*

Centre opencast infill
D7/013b (Late Bronze Age?): *Salix sp.* (willow) twig fragments, *Betula sp., Rubus sp* (bramble), *Juncus sp.* (80+ fragments), buds and scales
D7/013.1 (Bronze Age): *Salix sp.* twigs and leaf, *Quercus sp.* leaf, *V. vitis-idea* leaf (cowberry), *Betula sp.* (23 fragments), *Viola sp.* (violet), *Calluna vulgaris* (22), *Vaccinum myrtilus* (bilberry-24), *cf. Jasione montana* (sheep's-bit), *Juncus sp.*, Poaceae >2mm (grasses - 47), Poaceae <2mm (grasses - 10), bud scales etc., moss.
D7/013b: unidentified lichen on branchwood

Mine entrance A (Early Bronze Age mine sediments)
D7/047 (wood from launder): *Alnus glutinosa* (alder)
D7/049: Twigs of *Corylus avellana* (hazel), *Quercus sp.* leaf, fern leaves, plant fibres, scales and moss
D7/049b: 11 pieces of twig - *Quercus sp* (4) and *Corylus avellana* (3)
D7/050: 12 twigs - *Corylus avellana* (9) and *Salix sp* (3). *Alnus glutinosa* and *Quercus sp.* branchwood. Plant remains incl. *Fragaria vesca* (wild strawberry), *Juncus sp.* (rushes), *Luzula sp.* (wood-rushes) and moss
D8/050 (A/C section 3): moss probably *Calliergon cuspidatum*
D8/051: unidentified fragment of fern
D7/052: 3 small twigs of *Corylus avellana*. Plant remains incl. *Hypericum sp.* (St. John's wort), *Potentilla erecta* (tormentil), *Juncus sp.*, *Luzula sp.* (wood-rushes), *Carex spp.* (sedges), moss
D8/053 (A/D): 2 pieces of wood identified as *Corylus avellana*

Figure 62: Fragment of fern (bracken frond) as example of contemporary vegetation preserved in ochreous clay within EBA mining levels (D8 A/D 050). Photograph ST 1995.

Discussion

The peat horizon with remains of birch, willow, oak and hazel covering the floor of the mine gallery (D3/019)suggests an accumulation of wind-blown or washed-in leaves fallen from trees growing around the margins of the opencast during the Late Bronze Age.

Most of the seeds recorded from the base of the mine infill, or from the mining sediments themselves, notably heather, bilberry and tormentil, are typical of an upland moorland environment. Within the immediate vicinity damp ground conditions are indicated by the presence of rushes and sedges, and by moss inside the opencast. The results are in agreement with the pollen evidence from the site which reflects the development of blanket peat from the Early Bronze age onwards. However, leaves, seeds and scales of *Betula* suggest birch woodland growing locally, although it is perhaps worth noting that these seeds are well designed for transport by the wind and therefore the birch may have been a little distance from the site, but still relatively close. If the leaves, small twigs and wood fragments of hazel, willow, alder and oak represent natural accumulations, then it suggests that these species were growing within the immediate area of the site. Alternatively, if this represents material deliberately collected for fuel and mine timber etc. then these could be from much further away. If this was the case, there is no indication that the leaves were first removed. Overall, the results complement the pollen analyses from the site which indicate that oak and hazel were the most abundant tree species within the general area (Mighall and Chambers, 1993).

NOTE (by S.Timberlake):

Several broken hazel nut shells associated with cut *Corylus* branchwood were recovered from the mining sediments lying within the base of the entrance to the mine (D8/050-055). The fracturing of these shells was consistent with squirrel damage although the use and collection of these by the miners remains a possibility. More particularly, the evidence would seem to suggest that at least some of the wood may have been gathered in the Autumn months.

BEETLES (Sarah Clark, University of Sheffield)

Soil samples ranging between 0.25 and 5 litres were taken from five contexts within the mine infill sediment. They were processed for insect remains using the standard paraffin (kerosene) flotation method originally described by Coope and Osborne (1968). Beetles were identified using keys and the reference collections in Doncaster Museum and Department of Archaeology and Prehistory at the University of Sheffiled, taxonomy follows Lucht (1987).

The oldest context analysed for sub-fossil beetle remains was D7/013b (Middle-Late Bronze Age) within the main mine infill section. Species such as *Rhamphus pulicarius*(Hbst.), *Phyllobius maculicornis*(Germ.) and *Phynchites tomentosusi*(Gyll.) indicate the proximity of moist deciduous woodland comprising of Saliaceae (willow family) and *Betula* sp. (birch). Other beetles indicate developing blanket mire, e.g. *Olophrum piceum*(Gyll.) and *Acidota cruentata*(Mann.), with stagnant bog pools populated by species such as *Agabus bipustulatus*(L.) and *Hydrobius fuscipes*(L.) (Clark 1997).

Overlying this context was 013.1b where the sub-fossil fauna are dominated by woodland biota, including *Bryaxis curtisi*(Leach), *Rhagonycha lignosa*(Müll.) and *Cerylon histeroides*(F.). Most of the species present again favour Betulaceae and Saliaceae, e.g. *Deporaus betulae*(L.), but some live on *Quercus* spp., e.g. *Rhynchites aenevirens*(Marsh). This is consistent with the pollen record (Mighall & Chambers 1993). Species associated with grassland, e.g. *Phyllopertha horticola*(L.) and dung beetles, e.g. *Aphodius* spp. indicate nearby grazing by large herbivores. Wetland species and those associated with heather, including *Micrelus ericae*(Gyll.) and *Strophosoma sus*(Steph.), reflect the increase in mire community taxa shown by the pollen record (Mighall & Chambers 1993).

The insect fauna from the Late Bronze Age horizon 019 overlying the floor of the mine gallery (D3) are not dominated by any one biota, but show significant levels of woodland species correlating with the upper part of the pollen profile (Mighall & Chambers 1993). The diversity of species suggests an increase in moist, deciduous woodland with some *Quercus*. The pollen record shows an increase in grassland taxa (Mighall & Chambers 1993), but this is not mirrored by the insect fauna which is perhaps indicating more local conditions immediately surrounding the opencast.

Back within horizon 012b (Iron Age?) shows a dramatic decrease in woodland species. The majority of beetles retrieved (>60%) are aquatic especially *Hydraena britteni*(Joy) which is often found in bog pools. There are also a variety of species associated with grassland and heathland. This is consistent with the large amount of Poaceae (grass) pollen recovered (Mighall & Chambers 1993). The sample appears to reflect a landscape not unlike that of the present day, a well grazed permanent grassland drained by streams and blanket mire with bog pools.

Nineteen of the species retrieved from the early medieval turf layer (D3/009), were typical of rough unimproved grassland similar to that pertaining today, including at least 180 specimens of *Phyllobius roboretanus*(Gred.).

Woodland species comprised less than 10% of the sample suggesting a similar landscape to that reflected by horizon 012b. The pollen record also showed a decline in woodland taxa (Mighall & Chambers 1993). Large numbers of wetland/aquatic species were retrieved indicating the wetland insect community was flourishing on a blanket mire unaffected by contemporary land use.

Figure 63: Beetle habitat groups represented within peat infill deposits of prehistoric mine (from Clark 1996-97)

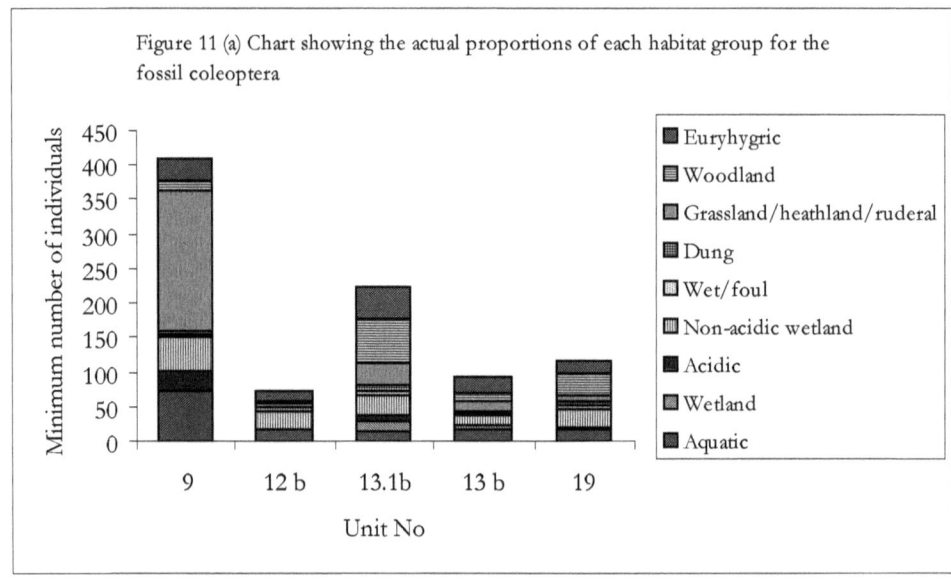

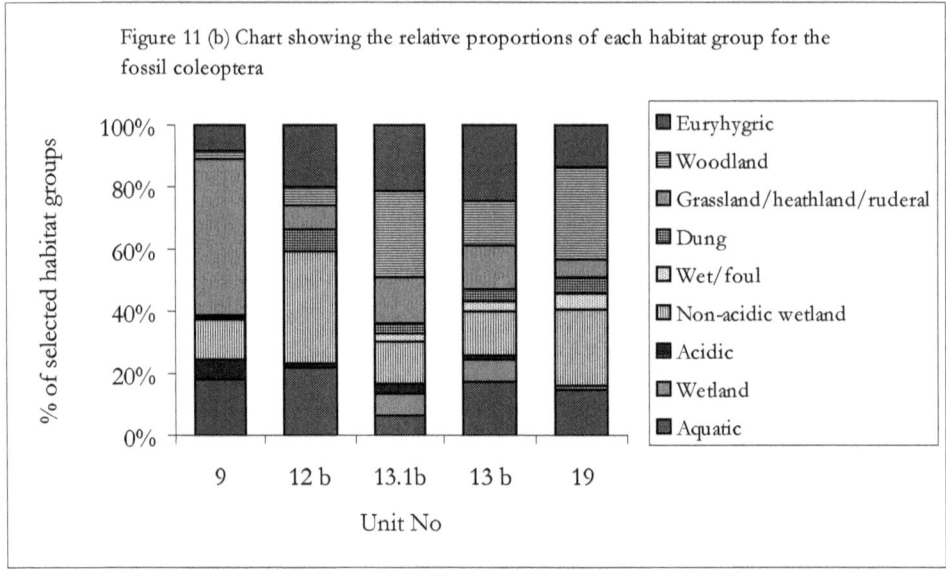

WOOD AND CHARCOAL (Su Johnson and Nigel Nayling, University of Wales, Lampeter)

The material examined comprised waterlogged and/or charred wooden objects held in the organics store of the NMGW, plus samples encountered during the excavations but not retained. From these thin sections were prepared. The tree-ring counts and growth-rates of a subset of these samples were recorded where preservation allowed recovery of this information.

Timbers, which were oak *(Quercus* spp.) and retained more than 50 annual rings, were sampled for dendrochronology at the Lampeter Dendrochronology Laboratory following the methods described in English Heritage (1998). The complete sequences of growth rings in the samples were measured to an accuracy of 0.01 mm using a microcomputer based travelling stage (Tyers 1999). All the measured sequences from this assemblage were compared with each other and with a range of external reference chronologies.

Figure 64: Tree species/genus percentages for the total wood assemblage

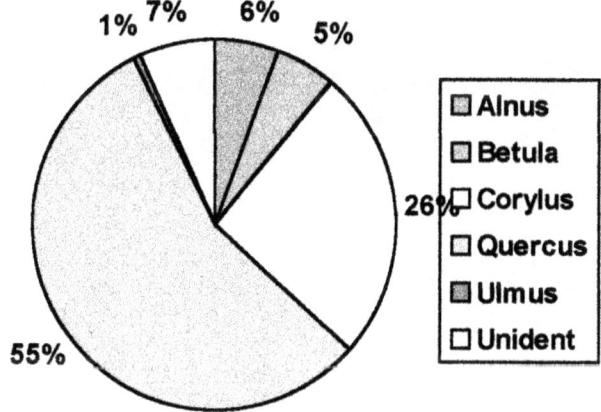

Waterlogged wood
The range of species encountered in the wood assemblage is relatively restricted comprising alder *(Alnus glutinosa),* birch *(Betula* spp.), hazel *(Corylus avellana),* oak *(Quercus* spp.), and a single identification of elm *(Ulmus* spp.). Wood identification of thin sections from a total of 136 samples was attempted and the results are summarised in Figure 63, indicating the dominance of oak (the majority of which is derived from context 081) but also the use of a significant amount of hazel, alder and birch. The poor condition of some of the assemblage is reflected in the 5% of samples that could not be identified.

Charred wood
The only species identified from amongst the 22 samples of charcoal examined from mine spoil were of oak. A number of these could not be identified - usually because the charcoal was highly fragmented.

A total of 17 pieces of waterlogged wood were recorded as having been partially charred. With the exception of single identifications of alder, birch and hazel, all identified, charred samples were also of oak. Ring counts from four of these oak samples range from 22 (incomplete) to 49 years with calculated average ring widths ranging from 0.63mm to 1.59mm. It would be inappropriate to infer too much from such as small sample, but the data would seem to suggest preferential selection of oak for fire setting in the mining operations. Tree-ring results would be consistent with the use of relatively slow-grown scrub oak branchwood for this purpose.

NOTE (S.Timberlake):

Most of the charcoal samples identified at the Dept. of Scientific Research at the British Museum (pers corn. C.Cartwright) in 1990 were similarly of oak, although one piece from the external tips was of pine *(Pinus sp.)* and another from inside the mine was identified as *Leguminosae* (probably gorse or vetch). A still earlier identification of mixed ash, hazel and oak charcoal from the external mine tips (Timberlake & Switsur 1988) has not been repeated, and now appears suspect. It seems likely that most of this was also oak.

Launders

The launder (083 (CH99:wl)) found within the base of the worked-out Vein Fissure 1 was hewn from an exceedingly slow grown oak. Attempts to measure the tree-ring sequence proved unsuccessful as the rings were so narrow that ring discrimination became unreliable. The parent tree must have been at least a hundred years old with continuous slow growth of less than 0.5mm per year. The other two launders examined were both of alder.

Context 013

Four samples from this peaty infill deposit were examined. Two of these samples were oak with diameters of 35mm and 65mm aged 17+ years and 45 years respectively. Both pieces have narrow rings (average ring widths of 0.88mm and 0.69mm) consistent with the wood coming from the branches of scrub oak. Charring of one of these (sample 16a) suggests that the deposit is not purely natural in origin.

Context 081

A total of 67 pieces of waterlogged wood are associated with this part-natural/ part-mining context. Oak roundwood (43 samples) dominates this group. The condition of the oak precluded confident identification of the season of 'felling' but tree-ring studies of this subgroup indicate that the majority is derived from roundwood stems 17-23 years old with moderate growth rates of between 1.4 mm and 2.4 mm per year.

Outliers to this distribution include a piece of pointed and burnt wood (1 02) with 80 rings (felled in the summer) and a partially charred piece of half-split oak with 43 surviving heartwood rings. The second most frequently identified species, hazel, comprises twelve samples of roundwood, derived from bulk sample 1.53. Six individually collected samples of hazel roundwood had diameters in the range 22-52 nun. Birch (6), alder (1), and elm (1) form a minor component. Again, these show no indications of intentional conversion.

Context 053

A total of 19 samples or objects from this 'fine grained mine sediment' encountered in the entrance area beneath the launder were examined. This group included single occurrences of charred stems of alder and birch, seven pieces of hazel brushwood with an average diameter of 8mm, a piece of larger diameter (43mm) hazel roundwood, pieces of charred oak roundwood examined in store, and nine samples of oak roundwood. The latter subgroup includes 3-4 year old stems, generally 6-7mm in diameter but also a 27-year old stem with a diameter of 14mm. Again, the number of samples in this
subgroup is small but the relatively slow growth rate of most of the stems suggest branch-wood from mature trees rather than coppice stems. Five of these samples exhibited partial growth of the last ring indicating 'felling' after April when wood growth commences, during the summer.

Context 054155

This group is dominated by small diameter roundwood with seven of the ten samples identified as hazel. This includes basket 079 and associated withies but also a charred artefact (the 'fire-stick' (CH96:wl)). Other items include a chip of oak from woodworking. The compressed state of much of this material precluded tree-ring studies.

Dendochronology

Despite detailed examination of all the oak assemblage, only three pieces were identified which had sufficient rings for analysis. One of these (from the oak launder) was so slow grown that the rings could not be distinguished from one another. Of the other two measured samples (from the 'stemple' and a burnt point from context 081), no significant computer correlations and visual matches were identified, either between the two sequences or with externally dated chronologies.

Figure 65: Hazel twigs strewn over and stamped-in to the surface of a working floor within the mine entrance (A" D8 Entrance A, Layer 053). Photo ST 1995

Figure 66a: Scatter diagram of ring counts and average ring widths (mm) for oak samples from 081

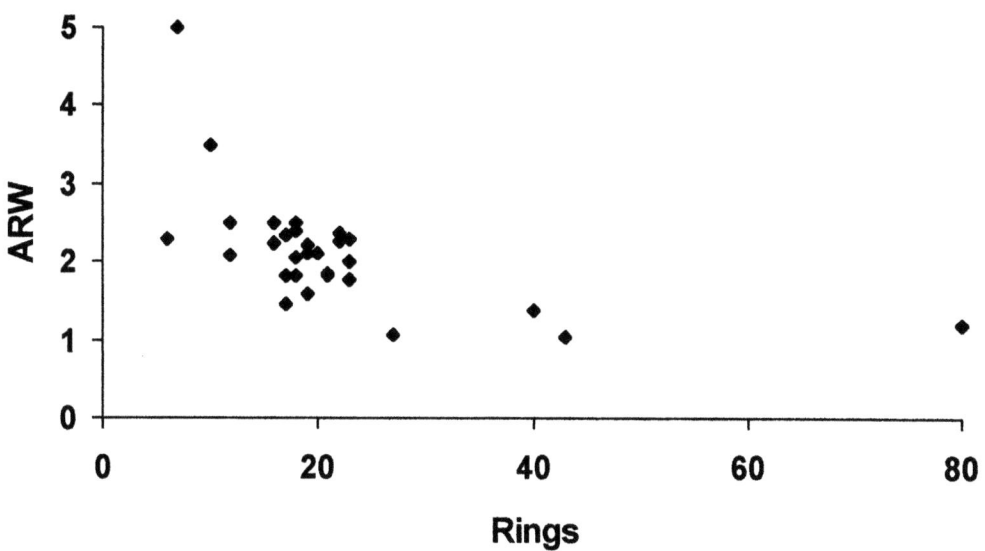

Figure 66b: Scatter diagram of ring counts and average ring widths (mm) for oak samples from 053

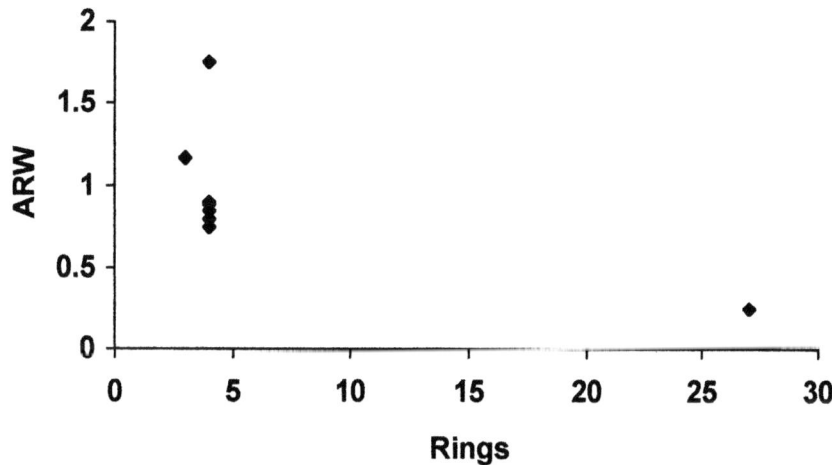

CHAPTER 10

THE PALAEO-ENVI[RONMENTAL EVIDENCE (PALYNOLOGY AND PEAT GEOCHEMISTRY)
(Tim Mighall, Coventry University)

POLLEN ANALYSIS

Four peat monoliths and cores were taken from the dissected blanket peat covering the plateau of Copa Hill above and to the north of the mine, less than 1.5 km from the excavations. The nearest sample site (CH2) was approximately 600 metres NNE of the opencast (SEE Copa Hill survey in Fig. 10). For comparison, organic sediments were also sampled from within the opencast itself, with pollen extracted from contexts both contemporary with and post-dating the Bronze Age mining activity.

Sub-samples of 0.5 to 1 cm thickness and two grains wet weight were prepared for pollen and charcoal analysis using the procedure described by Barber (1976) from one of the monoliths, CH2, to investigate the possible impact of prehistoric mining on vegetation. A detailed discussion of the pollen data was provided by Mighall and Chambers (1993). Some of that data is provided in Figure along with additional pollen data from the postprehistoric mining period to the present. A basal radiocarbon date obtained from the CH2 peat monolith (at between 134 to 133 cm depth) places the time of peat initiation at 3470 $\pm$ 35 years BP. A second radiocarbon date of 2395 $\pm$ 35 years BP was derived from a sample between 105 and 106 cm. Thus, the known period of Bronze Age mining corresponds in age with the vegetational record between 134 and 126 cm. A series of short-lived declines in the tree and shrub pollen are characteristic of the basal 30 cm. A noticeable drop in arboreal pollen occurs during the period of prehistoric mining, especially of *Quercus* and *Corylus*avellana-type. Non-arboreal pollen taxa normally associated with human activity, including Cereal-type and *Plantago lanceolata*, are also present within the pollen record.

The interpretation of the palynological data from the hilltop peat at Copa Hill is perhaps constrained by the geographical location and topographical difference of the pollen sampling site in relation to the prehistoric none. For instance, it is conceivable that timber used by prehistoric miners was cut from woodland growing on the valley sides and floor, both of which are some distance from the pollen sampling location, or, alternatively, that the regional arboreal signal was sufficiently strong to buffer the pollen record against a local reduction in arboreal pollen. In either scenario, the impact of any significant reduction in woodland has not been recorded within the pollen record captured by the hilltop peat. Nevertheless, a preliminary re-investigation of the site using fine resolution pollen analysis (sample thickness reduced to every 2 or 3mm; Jones, 2001) suggests that the original conclusions made by Mighall and Chambers (1993), linking small-scale declines in pollen with mining, still remain valid.

Examination of the Copa Hill mine sediments provided another opportunity to ascertain the true extent of the impact of mining on local woodland. High numbers of arboreal pollen were found within these, and provided some evidence for the presence and exploitation of mixed woodland, dominated by oak and hazel. In particular, *Quercus, Corylus avellana*-type, *Ainus, Betula and Ulmus* pollen is found in stratigraphic units (013f, 036, 046, 050, 053, and 054) within all three excavated areas, yet this gave no indication of the extent or changes in woodland structure that took place. Nevertheless, the total arboreal pollen percentages within these stratigraphic units, which yielded a relatively high pollen count, were slightly higher when compared to those in the hilltop peat, suggesting that a sizeable proportion of any woodland was confined to the valley sides and floor.

Whilst the palynological evidence supports the plant macrofossil data for the use of oak and hazel within the mining operations, the post-mining sediment, for example that from contexts 058 and 013f, also contains relatively high numbers of arboreal pollen, suggesting that significant woodland persisted in the Ystwyth Valley well after the cessation of Bronze Age mining. The permanent loss of woodland from the Ystwyth Valley occurred later on during the Bronze Age and possibly into the Iron Age. Two phases of woodland clearance can be observed in stratigraphic unit 0 1 3b, as arboreal pollen percentage falls from 30 to 25 cm and then from 1 5 to 1 0 cm. Similar types of vegetation changes are also characteristic of other parts of upland Wales, suggesting that mining here may have had no more than a small-scale effect on an otherwise common trend.

Permanent removal of woodland cover, the expansion of grasslands and peat development was taking place throughout the uplands of raid and south Wales during the late Holocene (Caseldine 1990). The two closest site to the Ystwyth are at the head of the Elan valley to the east and Plynfimmon to the north. Moore and Chater (1969) suggest that phases of woodland clearance and regeneration occurred during the Bronze Age at each site but there are no radiocarbon dates for either profile. At Cefn Gwemffrwd, peat expanded over areas cleared of woodland during the Bronze Age between 3335 $\pm$ 80 and c. 2755 years BP (Chambers, 1982a). Wiltshire and Moore (1983) attribute major changes in vegetation at Pwll-nant-ddu in central Wales to the activity of Bronze Age cultures. Peat development has also been linked to woodland clearance and agriculture during the Bronze Age at Cefn Ffordd (Chambers, 1982b). In South Wales, Bronze Age peoples are linked to the loss of tree cover at Coed Taf around 3000 BP (Chambers, 1983) and woodland disturbance at Waun Fach South and Gader

Fawr in the Black Mountains (Price and Moore, 1984; Moore et al., 1984). Open grass-heath moor and blanket peat formation also occurred in the Cleddau valley during this period (Seymour, 1985).

MINE SEDIMENT AND PEAT GEOCHEMISTRY

Full details of the results of a geochemical study of the blanket peat that occupies the northern plateau of the Ystwyth valley have been published elsewhere (Mighall et al., 2000a, 2000b, 2002a, 2002b) whilst results of palaecoecological and geochemistry analysis of some of the mine sediments is provided in Mighall et al. (2002c). The main findings are briefly summarised here to support the interpretation of the archaeological excavation. To reconstruct the pollution history of the prehistoric mine three peat monoliths were extracted from the closest, most suitable location, 600 metres to the north of the mine. A fourth monolith was recovered from a freshly exposed section approximately 1.35 km NE from the prehistoric mine to act as a spatial control. The peat monoliths were taken from freshly exposed sections of blanket peat and stored in either 15cm x 15cm aluminium monolith tins or plastic tubing. They were then wrapped in plastic, sealed and stored in a cold store. Chemical analysis by acid digestion ($HNO_3 - HClO_4 - H_2SO_4$) and atomic absorption spectrophotometry (AAS) following the procedure outlined in detail by Foster et al. (1987) or ICP-MS was carried out a contiguous samples. Two of the monoliths were radiocarbon-dated. Radiocarbon dating of the basal sample from one site closest to the mine, CH2, provided a uncalibrated date of 3470 ± 35 years BP whilst the sample extracted from between 105-104 cm provided an uncalibrated date of 2395 ± 35 years BP. These dates enclose the section of blanket peat that accumulated whilst ore was being mined (and following its abandonment) on Copa Hill. Two of these chemical profiles are shown in Fig. 66a (for CH2). The radiocarbon dates of this would seem to confirm that copper enrichment occurred in the peat during the known period of prehistoric mining. Similar enrichment of copper was evident in an adjacent profile and one taken 30m away (See Mighall et al. 2002a). One radiocarbon date of 4240 ± 70 years BP was determined from a sample taken from 178 to 180 cm at site furthest away from the mine, CH3 (see Fig.66 b). Here, copper was not enriched during the known period of prehistoric mining. Conversely, lead concentrations remain low (below 55 μg g^{-1}) within the basal sections of every monolith (Fig. 66 b), although there are slight increases at the very base of some of these profiles. Lead enrichment is first noted within the section of CH2 believed to correlate with the period of the Roman occupation whilst both lead and zinc concentrations increase from the Medieval period until the early part of 20th century (see Mighall et al. 2002b). Similar enrichment of lead and zinc is shown in the remaining profiles. Whilst other possible explanations are discussed, it is argued that the high lead concentrations represent evidence of atmospheric pollution caused by mining. Zinc, however, may have suffered from post-depositional mobility.

Figure 67a: Pollen diagram and geochemical profile from core site CH2 (T.Mighall)

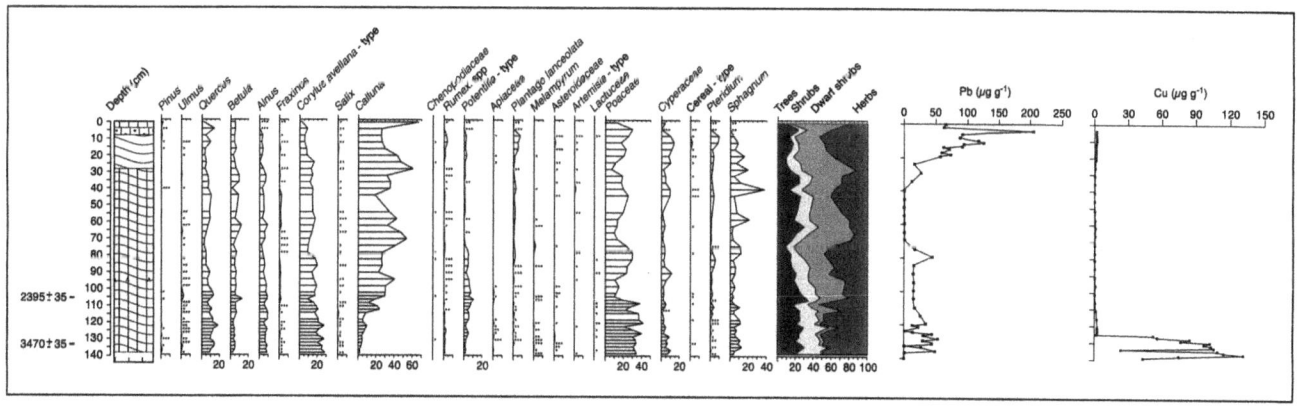

Figure 67b: Geochemical profiles for both lead and zinc from peat monoliths CH2 (**A and E**), CH4 (**B and F + C and G**), CH1 (**D and H**), and CH3 (**E and I**), PLUS coring site location map of moorland tract above mine. After Mighall et al. 2002b.

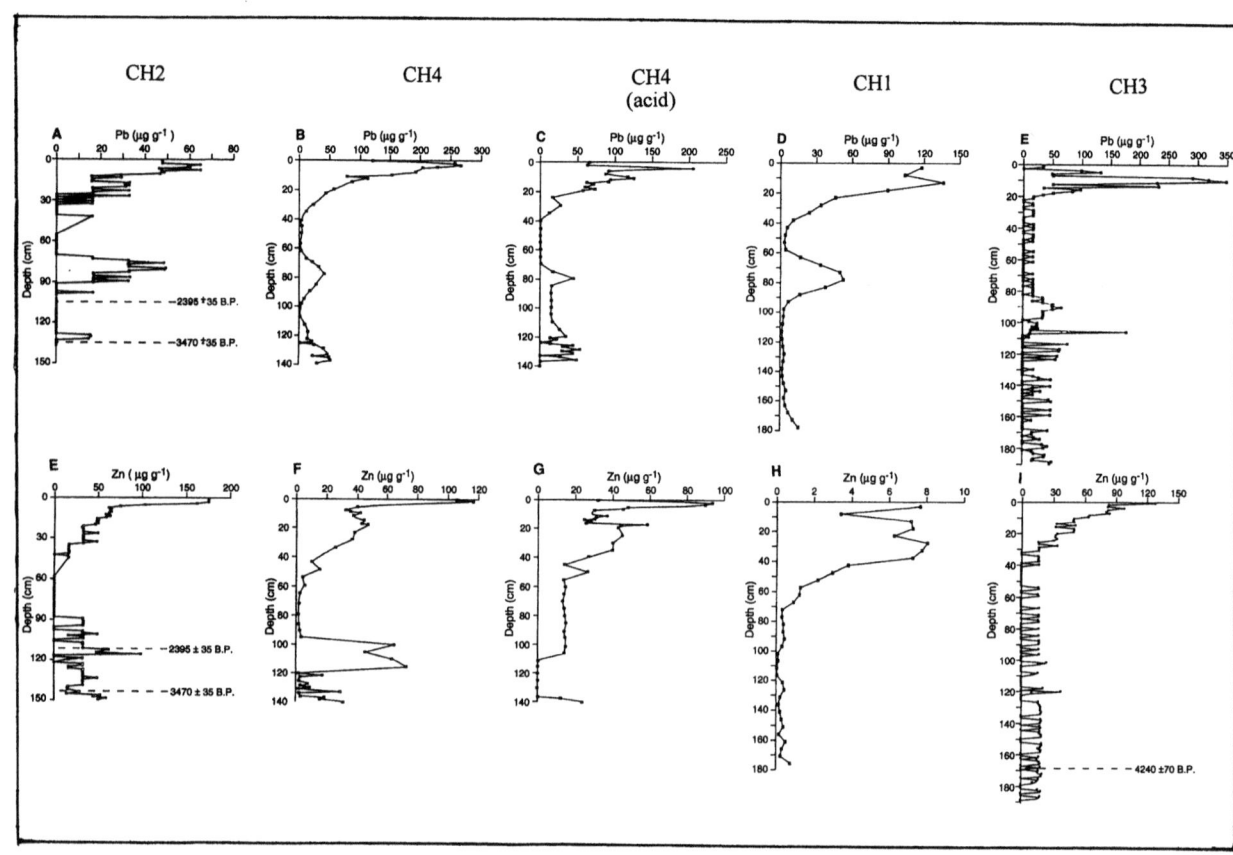

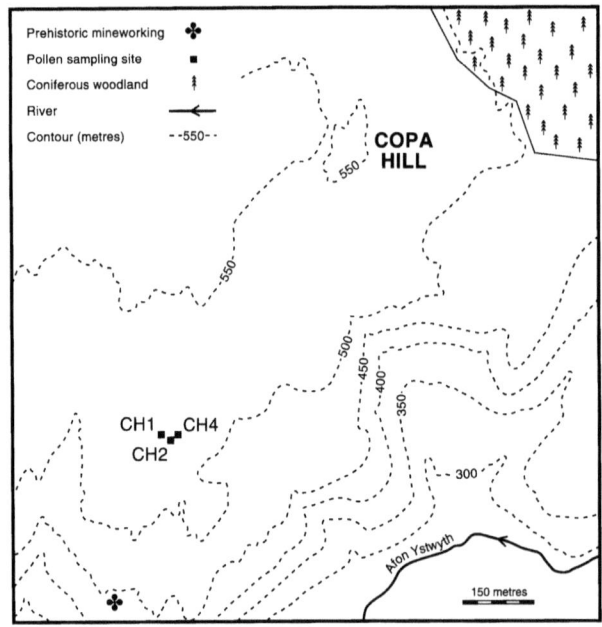

CHAPTER 11

WOODEN ARTEFACTS (S.Timberlake)

LAUNDERS AND THEIR FUNCTION

The discovery of three wooden launders within the mine, plus the possibility of still others lying discarded within the as yet un-excavated bottom of the opencast, raises the question as to whether these once formed part of a connected series of open pipes acting as an aqueduct, channelling water across or around the sides of the working. More realistically, these would have formed part of a drain to help remove excess water from the mine - either directly from the point of entry such as a spring, or else from the place where it was baled into the launder by hand. The *in situ* survival of the 4.7 m long alder launder (047), in a position and at a gradient which would clearly have removed water from the mine, supports the notion that this was for drainage. However, the narrowness of the opening (an internal diameter of between 14-18 cm) presents something of a problem here, since it would never have been easy to bale water into, instead this suggests the careful pouring in of water at one end, perhaps from a waterfilled bag or skin raised from a drainage sump within the bottom of the mine, or else from the directed flow of a spring, skilfully tapped upon the wall of the working . If the latter situation was the case, then one of the discarded launders, such as the wider 1.5 + metre long oak section (083), may once have carried water from a possible ingress point (such as the outcrop of the SW-NE fault above the Deep Fissure (1)) over the opencut and then onto the long alder culvert located within the entrance cutting. In turn, this water may have flowed onto another (024?) located at the southern end of the entrance, the latter finally removing it from the mine. A function not previously suggested for these launder(s) may have been for the washing of hand-crushed ore clean of clay and dross, and perhaps also for carrying out simple waterborne gravity separation of the closely associated copper and lead sulphides (each with differing specific gravities), the lighter copper ore being washed further along the trough. It seems that this practice of washing and concentrating the ore, although not essential given the more time consuming fine crushing and hand separation / winnowing methods also available, was however being carried out from prehistoric times onwards, examples being cited by both Lewis (1996) and Ottaway & Wager (2000) at the spring sites('washeries') of Ffynnon Galchog and Rufeiniog close to the Bronze Age mine on the Great Orme, also by Pittioni (1951) on the Mitterberg, and by Ambert et al. (1983) at Roque Fenestre, Cabrieres. Ore washing techniques had evolved considerably by the time of the exploitation of the Athenian silver mines at Laurion (Ellis Jones 1988), whilst during the Medieval period flat-bottomed or riffled wooden launders appear to have been in regular use for the separation of sulphide ores, as illustrated in the pages of Agricola (1556, 291, 348). However, the Copa Hill launders, with their narrow sides and round smooth bottoms would hardly have been ideal for this purpose (excepting perhaps the shallower oak specimen), yet it is conceivable that some or all of these may have had a dual role both in drainage and ore-washing, and as such these may have functioned as a moveable entity within the mine. Significantly, a careful excavation carried out at the National Museum of Wales in 1996 of a removed section of the Copa Hill launder (047 D) found no traces of washing residues or ore particles within the sediment that infilled it, although tooling marks upon the base of this were well preserved (*pers.comm.* A.Gwilt).

Analogous artefacts have been found within a number of other Bronze Age or early metal mines. Pittioni (1951, Pl.V) illustrates a 'wooden water pipe' up to 5 metres long (and approx 40 cm wide) excavated from one of the separating areas beneath the (Middle-Late Bronze Age?) mines at Kelchape in the Tyrol, whilst he also describes (p.51) the discovery of a 'wooden trough' 1.7 metres long and closed at both ends, the latter being interpreted as an implement for washing and used in conjunction with a sieve and wooden knives much like a buddle for separating and concentrating the ore. Thus two types of wooden trough and their functions appear quite distinct, the 'wooden pipe' being much more similar to the launders found on Copa Hill used for carrying a continuous flow of water, although the context and function of the former outside of the mine workings at Kelchape is not at all clear (Pittioni's suggestion being that it was 'used to divert sub-soil water from the mountainside'!). Bromehead (1950) refers to water being introduced *into* the mines (of the Mitterberg) 'by launders (troughs) of hollowed half tree-stems', implying a degree of misunderstanding over their function, whilst the context of recent discoveries made of by Clemence Eibner (*pers. comm.* B.Ottaway) are insufficiently well known. Nevertheless, at a Late Bronze Age smelting site within the Schwarz/Brixlegg area of North Tyrol it has been suggested that wooden launders were used for washing crushed slags in order to recover metal prills (Goldenberg & Rieser, 1995). In hindsight although it seems altogether more likely that closed wooden or pans rather than launders would have been more suitable for this purpose. The more common drainage implements found within primitive mines appear to have been wooden bowls or scoops, such as those found within pre-Columbian Indian copper workings on the Keewanaw Peninsula, Lake Superior, U.S.A.(Griffin 1961), adding weight to the idea that launders were meant to tap a continuous flow of water. Meanwhile, underground discoveries of hollowed-out log drainage launders have been reported from Late Bronze Age - Roman copper mines on Cyprus (Bruce et al. 1937), the latter as broken small diameter sections less than a metre long, whilst within excavations at the ancient copper mines of

Tonglushan, Hubei Province, in China the use of wooden launders dating from the 8th - 6th centuries BC have been described draining the underground workings to a sump at the base of a timbered shaft (Maddin 1986). More recently at Zawarmala in Rajasthan, India, an intact 3 metres long and 20 cm wide gutter has been found within the fireset stopes of an ancient silver/lead-zinc mine and this appears to be more than 2000 years old (HinduZinc 1989; Craddock 1995). The Zawarmala drainage launder, clearly intended for carrying water from areas of seepages to points where it could then be collected and removed of the mine, is perhaps the closest parallel we yet have for the examples from Copa Hill. There appear to be no records of anything similar from the British Isles, the only other hollowed log artefacts of similar age from Ceredigion being the poorly preserved remains of two primitive wooden coffins discovered beneath a burial cairn on Disgwylfa Fach, Ponterwyd (Savory 1980, 197; Briggs 1994).

These Early Bronze Age launders from Copa Hill are unique, and quite possibly the oldest examples of mine drainage equipment ever found.

Figure 68: Well-preserved section of launder remaining *in situ* within middle part of Entrance A. Photo ST 1995

Figure 69: Detail of (south) end (ST 1995)

Figure 70: View of same from SE side resting on stone supports within re-cut ditch in basal mine sediments. ST 1995

OTHER WOODEN ARTEFACTS

Apart from the launders some 75 small finds of worked wood were recovered from the mine, the great majority of which came from the prehistoric levels. Most appear to have been fragments of cut branchwood collected as fuel and for flooring material within the wet mine trenches, whilst less than 25% showed any indications of being recognisable artefacts. Amongst the latter was found possible evidence for stakes, planks, stemples, wedges, handles, (shovels?), ladder rungs, fire-sticks, lighting chips, withy hammer handles, rope and basketry - most of these indifferently preserved, and apparently discarded after breakage upon the rock floors and in mine spoil, although some objects appeared charred, perhaps after re-cycling as fuel. Organic preservation of the wood was particularly poor within some of the coarser void-filled sediments such as mine spoil and scree, the best preserved artefacts being recovered from amongst waterlogged layers of fine grained sand, silt and clay-rich sediment lying within puddles above the rock floors and ledges.

Wooden stakes.
The remains of some 12 wooden stakes or posts were recovered from the mine, none of them *in situ* and apparently all discarded as rubbish, most within the peat and scree layers that had accumulated following the abandonment of the working. Almost all were oak and most of these had been split (a few of as palings), with perhaps a few associated with intrusive features such as the medieval/postmedieval shafts encountered close to the north wall of the opencast. However, other examples were removed from Middle-Late Bronze Age peat horizons, including one complete post (CH90:w2) 99 cm long from area D3/027, and a well preserved point of a small roundwood stake (CH90:w11) recovered from the peat-covered floor of the mine gallery (D3/019). It seems likely that these may well have been thrown into the opencast from outside, and could originally have had an agricultural or domestic function, perhaps for fencing, tethering animals, or else as part of a shelter. The burnt end of a discarded oak stake (CH94:w6) was also excavated from amongst Early Bronze Age mining sediments(050). Here it had been re-used along with other branchwood to support the base of the launder (047).

Stemples, planks.
Except for one or two possible modern examples associated with the intrusive shafts, all of these were found deep within the workings and were associated with Early Bronze Age mining activity. Stemples for the most part consisted of substantial pieces of oak timber (with a minimum of 50 rings) of sufficient size to be jammed between the rock walls of an opencut and to take the weight of a man. Most were distinctive in that they had been cut to a wedge shaped termination at least at one end (a feature also typical of stemple timber within 'modern' mines) enabling these to be hammered down tightly against one of the rock walls, whilst the other end remained anchored, sitting within a rock-cut niche.

Examples of what were probably discarded and burnt stemples (but may have just been just substantial pieces of cut firewood) were also found within the base of the mine infill (Fig.76 B), whilst several large pieces of collapsed alder timber (e.g. 059b) may once have had a similar function, but more likely were parts of a ladderway, or supports for a launder, wooden staging, or else part a tripod for lowering or raising water or rock within the mine. No fragments of what could have been sections of 'notched tree trunk ladders', such as those recorded from other Bronze Age trench or fissure mines such as those at Derrrycarhoon, Co.Cork (O'Brien 1989) and Chinflon, Huelva, SW Spain (Rothenberg & Blanco-Freijeiro 1981; Andrews 1994) were recognised amongst this assemblage, all the substantial pieces of round timber being smooth and unmodified, except perhaps at their ends. Altogether, the preservation of these larger timber fragments was poor and no artefactual features could be identified. By far the best preserved example of a stemple, a 75 cm long square-cut oak section (CH99:w2 SEE Fig. 76 C) was recovered from the deepest part of the excavated area of the mine, a position into which it may have fallen from the open cleft of the narrow vein working (Fissure 1) above. Elsewhere in the mine a round oak beam stemple (CH99:w4 SEE Fig.76 A) was found jammed *in situ* in between the vein opencut walls in area D9, yet in both size and shape this was much less typical of a climbing stemple. In fact its thinness and relative elasticity suggests that it could have been used instead as a pivot over which a rope was slung to raise small bags or baskets from the workings below.

Comparison of the two wood assemblages found within the mines on Copa Hill and Mt.Gabriel in Co.Cork (Mines 3/4) shows that neither were strictly comparable. For example, no obvious examples of stemple wood were found at Mt.Gabriel, most of the larger pieces being sections of oak planking, reflecting the horizontal aspect of mining here and the need for steps or platforms set into the steeply inclined entrance footwall, yet with little or no vertical element such as is found within the deeper opencuts on Copa Hill (O'Brien 1994, 152). Of the few possible fragments of oak plank recovered from Cwmystwyth, too little remained to provide any sort of interpretation as to function. More common was the evidence for woodworking, particularly the shaping of stemples, which was being carried out on site. Wood shavings (mostly oak) were ubiquitous within the waterlogged mining sediments, whilst off-cuts, possibly from the ends of oak stemples were found discarded upon the floor of the entrance passage (e.g. 078 (CH95:w3) SEE Fig. 78 F). Stemples such as these may have been recovered from the mines of the Mitterberg, yet they are not specifically referred to by Pittioni (1951). However, several examples were removed from the excavated opencast at Chinflon, Huelva, SW Spain (Andrews 1994, 17), whilst amongst the workings of the Late Bronze Age - Roman mines in Cyprus such structural timber was commonplace , preserved and in some cases partially replaced by copper within this highly mineralized environment (Bruce et al. 1937).

Handles, wedges, fire-stick etc.
No recognisable fragments of wooden shovel were recovered from Copa Hill, although various unidentified whittled wooden objects, including what part of what appeared to be a broken spatula (CH91:w13) were found within of the base of the infilling scree/ mine spoil lying against the north wall of the opencast. This splitting and/or shaping of the ends of what were usually small pieces of oak roundwood gave some of these the appearance of being 'pegs' (CH91:w5 and CH91:w18), and others as fragments of solid wooden 'handles' or else 'ladder rungs' (CH91:w3 and CH96:w11). Although exact analogies for these have yet to be found, descriptions of wooden small finds from both the Mitterberg mines and Mt.Gabriel suggests that some of those objects with carved flat heads and points may have been examples of wedges or prise sticks (Pittioni 1951, Pl.VI; O'Brien 1994, 148), the more heavily whittled ones being pegs (O'Brien 1994, 151), whilst the carefully carved round sections may indeed have been handles, the latter perhaps from the ends of destroyed oak or alder shovels (O'Brien 1994, 146-147). Another 'handle' like fragment of shaped hazel roundwood (CH96:w10) was found within the basal sediments of the Mine Entrance (Fig.77 B). Alongside of this was found a 139 cm long 'fire stick' of cut hazel with a burnt and worn point, the latter possibly used as a poker for stoking the fire-setting hearths (CH96:w1; Fig. 77 A). No comparable example to this has been found at other Bronze Age mines.

Withy handle, ties, rope and basketry.
In 1995 the two broken halves of a twisted strand of stripped hazel (withy) were found lying close together on the floor of the Entrance (A). These were interpreted as the discarded fragments of a flexible withy handle formerly wrapped around the middle of a hammer-stone (Timberlake 1995). Although similar sized fragments of broken withy have resulted from recent experimental reconstructions and demonstrations in the use of mining hammers (*pers com.* B.Craddock), it should be noted that no intact hafted implements, or even any conclusive associations between hammer-stones and broken bindings have yet been recovered from an archaeologically excavated prehistoric mine. However, there is at least one reference to finds of original handles within the Mitterberg (Kyrle & Klose 1918), whilst the most celebrated discovery of all is undoubtedly that of 'Copper Man'. This 2000 year old mummified body was discovered in 1899 entombed along with several complete examples of his hafted stone mining tools and other implements in an old copper mine at Chuquicamata in Chile (Bird 1979). The method of hafting these stones has since been studied in detail on another example, and although this involved quite different materials, the basic approach using twisted withy type sticks bent and then tied off around the middle of the cobble head seems altogether quite similar. From Ireland, a twisted and looped hazel withy recovered from the interior of Mine 3 on Mt.Gabriel was also interpreted as a hammer handle (O'Brien 1994, 151-152), yet this shows much greater resemblance to what has since been described as withy rope or tie (CH96:w3; Fig.79 E) found within an adjacent horizon (055) of the mine entrance on Copa Hill. It has been suggested that the latter may in fact have been a 'bond' for a faggot i.e. it was used to tie up a bundle of firewood, a factor which would have made the handling of fuel and its carriage up to the site considerably easier (*pers com.* D.Goodburn). However, this rope tie was found intimately associated with various other fragmentary withy remains such as twisted and bent hazel rods, knots (CH96:w4 -5; Figs. 79 C&D), and a side panel or lid for a coarsely woven basket or wattle hurdle (CH96:w13; Fig. 80). The latter was apparently discarded on the mine floor following breakage and charring.

Figure 71: Two halves of broken withy handle (for hammer-stone?) excavated within mining sediments (053) in sector A/B of Entrance A. Photographed separately by ST 1995 (NB not in correct orientation SEE Fig.79)

Given their close location, these fragments, including the ties, could have come from a single disintegrated basket, suggesting the presence of associated flexible rope handles. A possible reconstruction of this artefact (Fig. 81) is based on the identification of two pairs of doubled stakes or sails about 8- 15 mm in diameter plus several plain weave rods of about 10 mm diameter. The latter appear to be twisted at the ends and wound round one of the sails, clearly showing this to be one end of a wicker panel or wattle hurdle. If this is a container, the method of construction suggests that this may have been part of a strong but shallow two-handled basket, perhaps one associated with a processing activity such as washing or else for gathering and sorting crushed ore (B.Craddock *pers.com.*).

Figure 72: Twisted withy tie associated with basket remains found in Entrance A A/C layer 055. Photo ST 1996

It has also been suggested that heavy creel-type baskets may have been used for carrying ore and rock out of the mine (D.Goodburn *pers.com*). The design of such a basket may be really quite simple - and indeed were commonly made in country areas up to the end of the 19th century winding. However, surviving examples of Bronze Age baskets are virtually unknown.

Moreover, there are no reports of discoveries of withy (hazel) rope or basketry at other British prehistoric mines. Many tiny fragments of hazel twig associated with hammerstones were recovered from the 1998 excavations of an Early Bronze Age pit at Engine Vein, Alderley Edge, Cheshire, yet it was not possible at the time to tell whether this was an artificial or natural accumulation (Timberlake 2003 *forthcoming*). Straw rope has been reported from the mines of Tonglushan, China (Maddin, 1986) whilst plaited (presumably hemp) rope was used within the Phoenician or early Roman mines of Cyprus (Bruce et al. 1937, 658), yet both these represent examples of more sophisticated mining technology.

Slightly more common are examples of stick or reed baskets used for carrying ore or waste rock, such as those found at Dariba in Rajasthan which date from the late 2nd - 1st millenium BC (Craddock 1995, 82) and from the aforementioned Cypriot mines (Bruce et al. 1937, 661), the latter consisting of a type of round basket with a much finer weave than the apparently crude example from Copa Hill. A still closer parallel for this type of coarse creel-type pack basket might be found in an example recovered during the 18th century from an ancient opencast trench at Esgairmwyn Mine some 7 km to the south-west of Copa Hill - the only record of this artefact being a crude thumbnail sketch (Morris 1752). However, it seems more likely now that that this find was medieval rather than prehistoric (Timberlake 2002).

Figure 73: Esgairmwyn basket (Lewis Morris 1752)

Although rarely preserved on Bronze Age sites it seems likely that the prehistoric use of hazel withies was once widespread in the manufacture of baskets, handles, rope and ties for all manner of different purposes, an example being two lengths of looped and knotted withy (virtually identical to those found on Copa Hill) from Goldcliff West on the Severn Estuary Levels, the latter believed to have been used either for structural purposes, such as tying hurdles, or else as parts of a portable artefact such as a bucket or fish basket (Bell 1993, 98). Twisted withies were used up until recent times by English woodsmen for tying faggots of firewood (Tabor 1994).

Figure 74: Example of traditional faggot bundle and withy tie

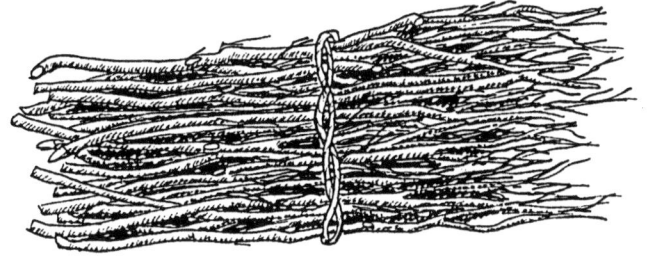

Figure 75a: **WOODEN ARTEFACTS -Launders from prehistoric mine** - longitudinal and cross-sectional profiles: **(A)** main alder launder [047] recovered from Entrance A (in sections), showing position of prominent tooling marks upon internal surface. Drawing B.Craddock

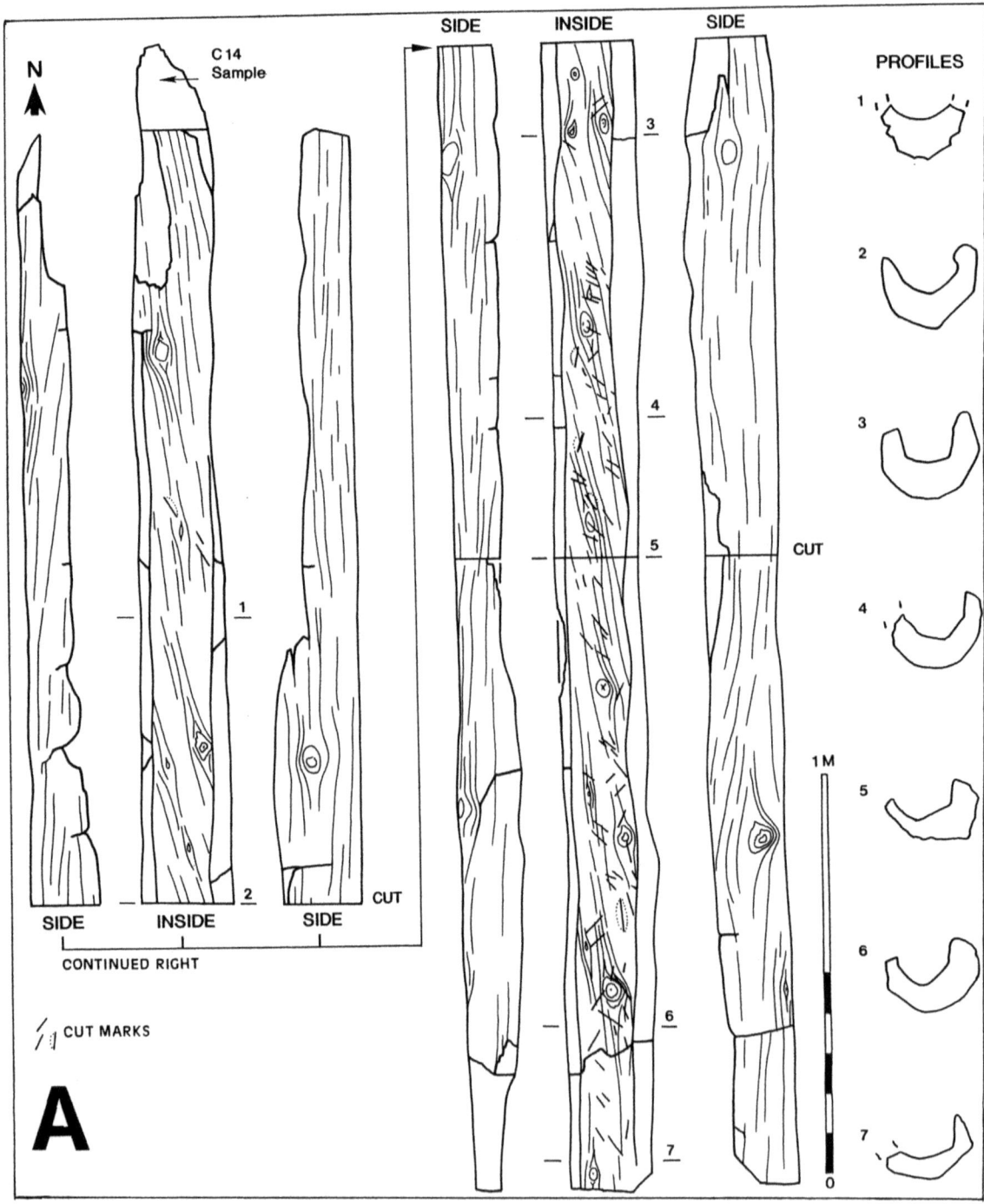

Figure 75b: WOODEN ARTEFACTS - Launder(s) from prehistoric mine (continued...)
(**B**) external view of fragment of damaged launder (024) discarded in and recovered from top of Deep Fissure 1; (**C**) large oak launder (083) recovered from bottom of Fissure 1. Drawings by B.Craddock 2001

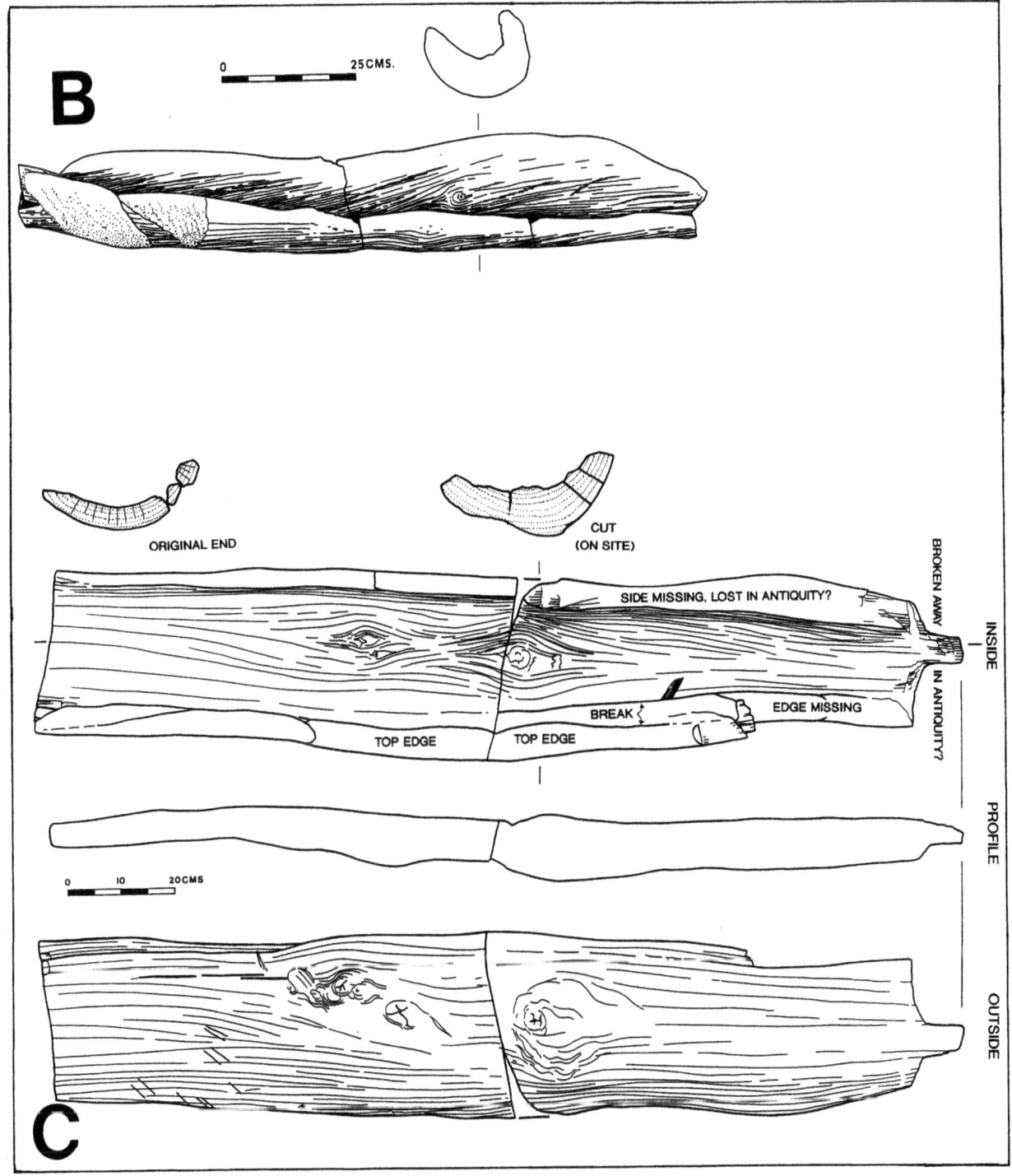

Figure 76: **WOODEN ARTEFACTS - Oak stemples and mine timbers** : (**A**) narrow beam stemple [CH99:w4 (103)] from D9 Vein Fissure 3; (**B**) end of burnt timber [CH99:w3 (102)] from base of Deep Fissure 1; (**C**) short square wooden stemple [CH99:w2 (101)] from base of Deep Fissure 1. Drawings B.Craddock 1999.

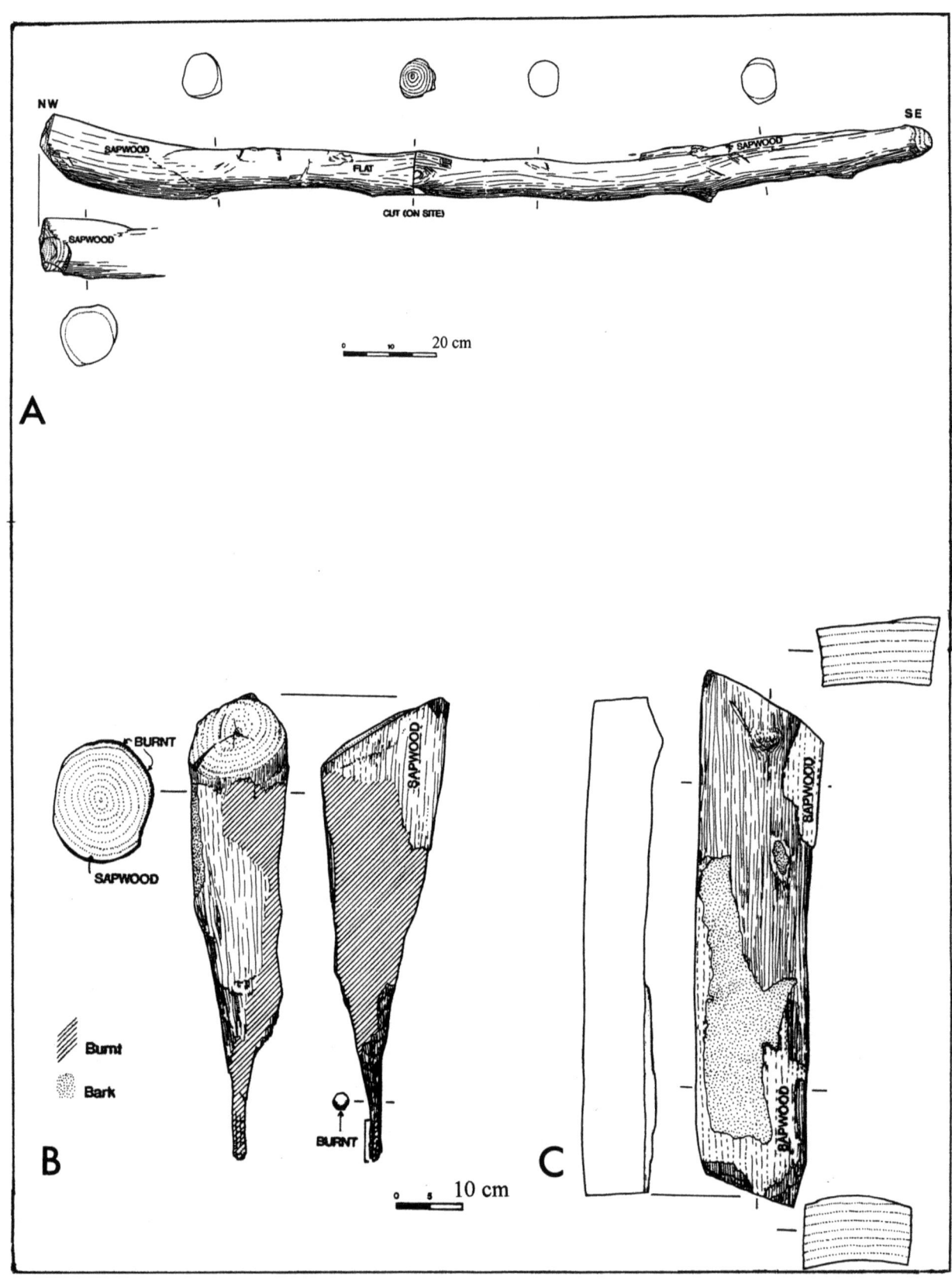

Figure 77: WOODEN ARTEFACTS - 'Handled' sticks and stakes: **(A)** fire-setting stick or poker [CH96:w1] from D8 A/C 055 with burnt and worn end; **(B)** cut and shaped/stripped piece of oak branchwood from D8 A/C 054/055 [CH96:w10] used as a rung or handle (?) N.B. chop marks; **(C)** small broken wooden stake [CH90:w11] found within Late Bronze Age peat layer (019) on floor of mine gallery D3; **(D)** rod of hazel with fresh-looking chop mark at one end. Found in Entrance A [CH95:w8]. Drawings by B.Craddock.

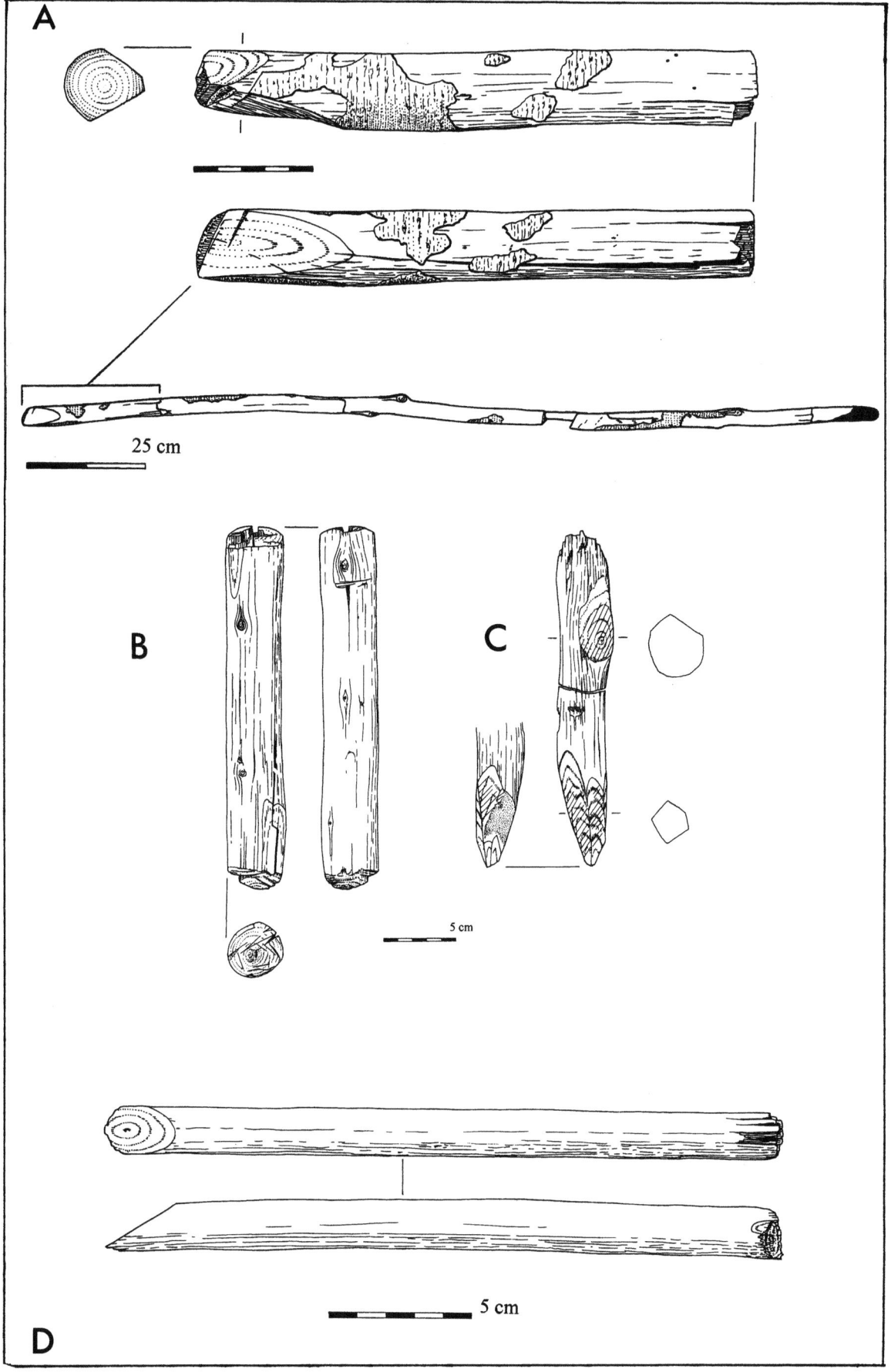

Figure 78: **WOODEN ARTEFACTS - oak chips or shavings**, possibly associated with the lighting of fires: **(A-D)** split wood chips from amongst mine spoil (layer 100) within base of Deep Fissure 1 [CH99:w6]; **(E)** small wood chip with well preserved fine axe blade cut at one end [CH99:w9], found amongst lowest organic infill (081) within bottom of Fissure 1; **(F)** off-cut end of oak timber [CH95:w3 (078)] from Entrance A A/D 054. Drawings B.Craddock.

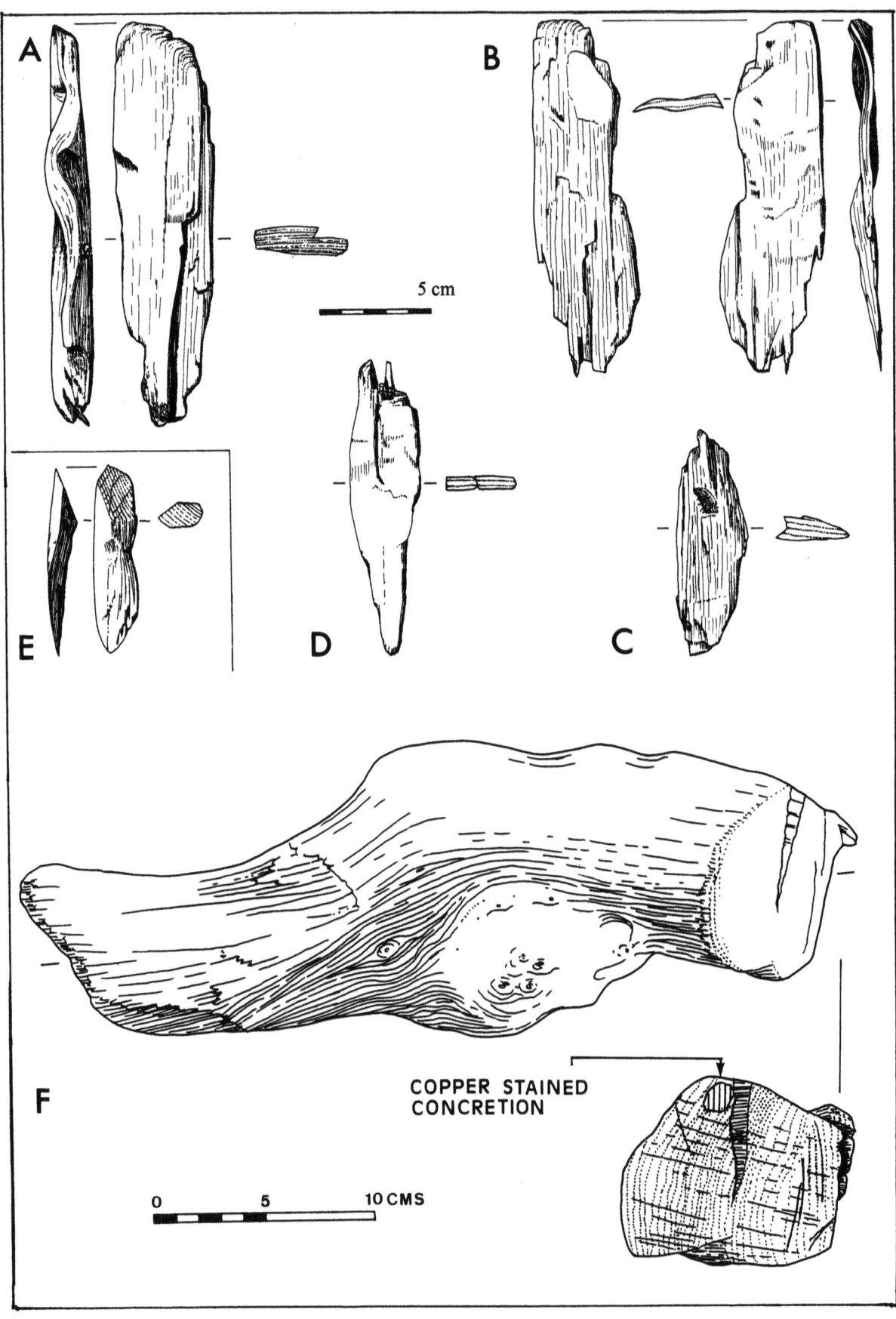

Figure 79: WOODEN ARTEFACTS - Twisted hazel withy handles, ties and rope: (**A**) part of handle (for a hammer or a wedge?) from D8 054; (**B**) a possible fragment of a coarse woven basket (from D8 A/C 054/055); (**C**) the end of a looped withy tie [CH96:w5] from D8 A/C 054/055; (**D**) a withy knot (part of a basket fastening) [CH96:w4] from same location; (**E**) a twisted withy tie or rope [CH96:w3], from same location; (**F**) the two halves of a discarded withy handle for a hammer-stone [CH95:w5 (076)] from D7/8 A/B 053. Drawings by B.Craddock.

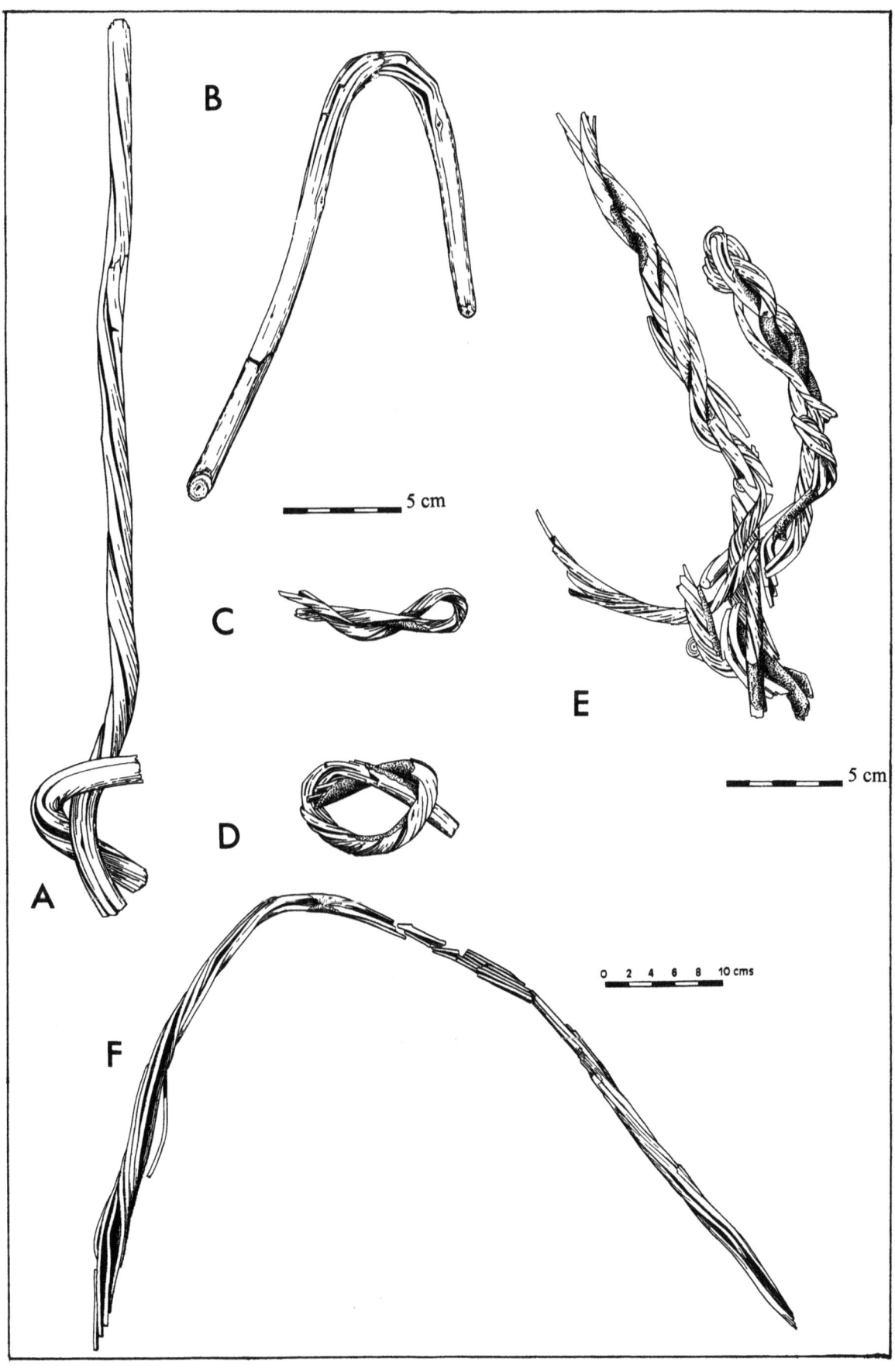

Figure 80 (TOP): **WOODEN ARTEFACT** - small fragment of coarsely woven **hazel basket** [CH96:w13 (079)] from D8 A/C 055 as drawn *in situ* by B.Craddock, 1996.

Figure 81 (BOTTOM): Suggested reconstruction of shallow basket (B.Craddock 2002)

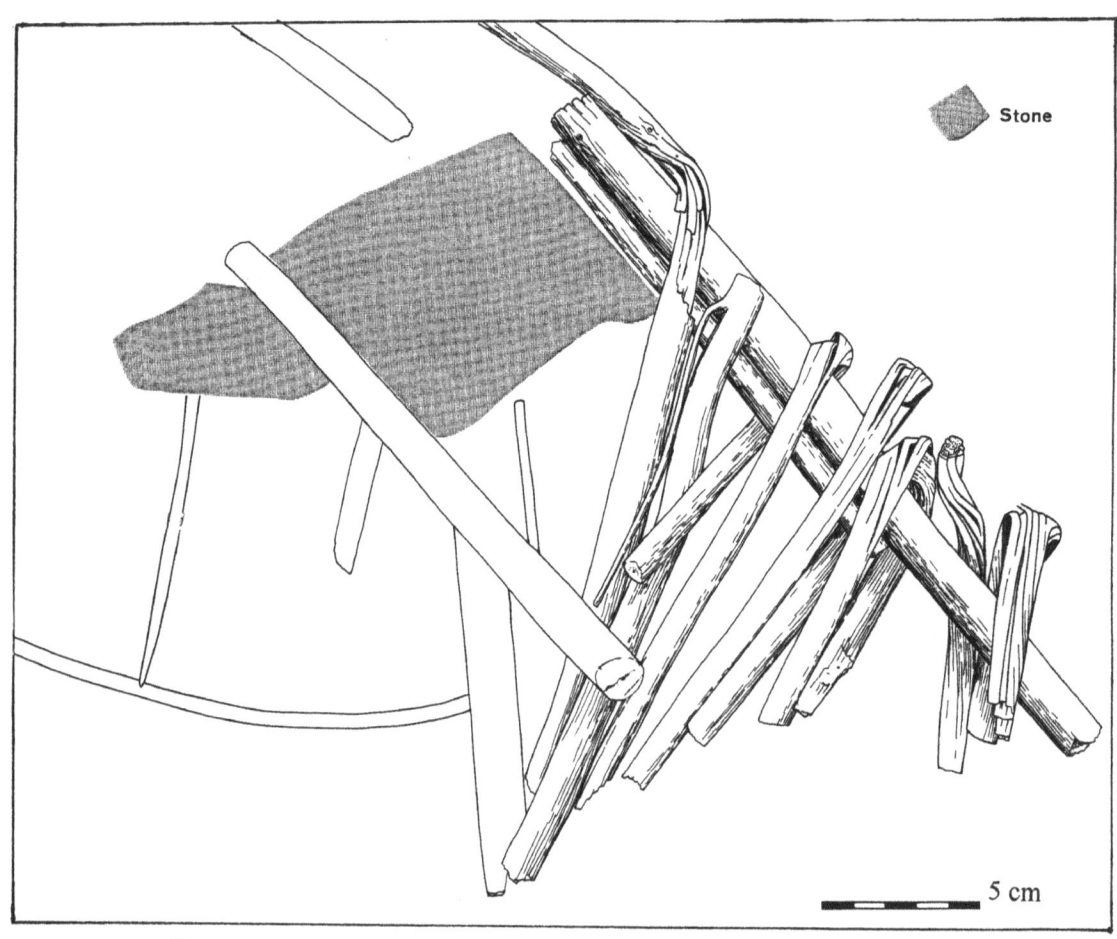

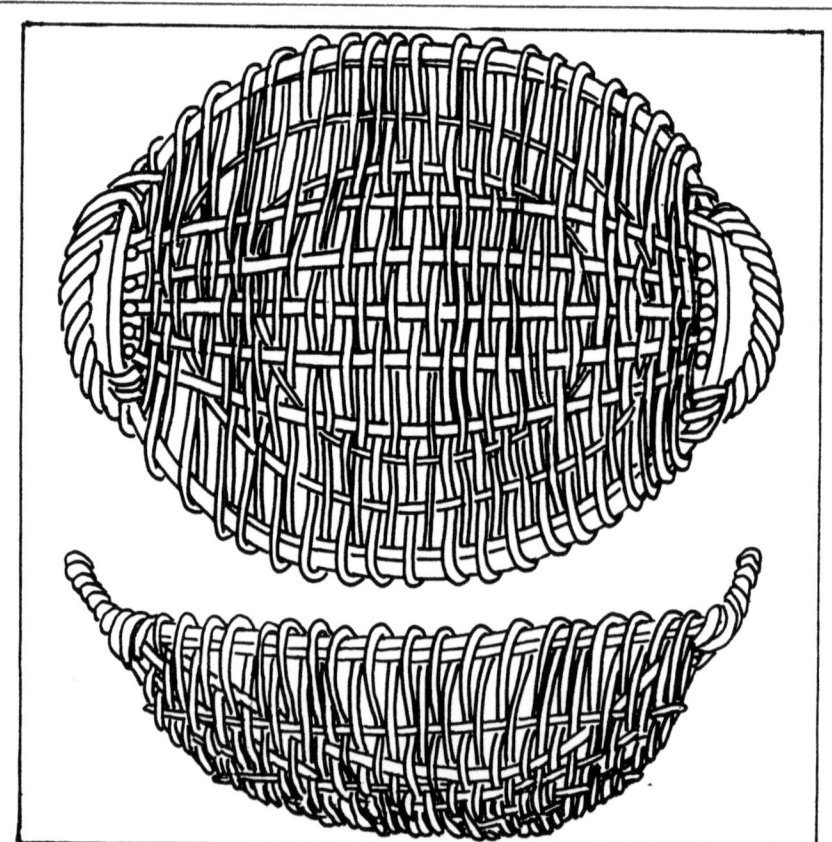

CHAPTER 12

WOOD TECHNOLOGY (Damian Goodburn)

A selection of items held in storage within the Department of Archaeology, National Museum of Wales, Cardiff were examined in February 1998. In particular, detailed notes and drawings were made of the alder launder (047) removed from the mine entrance.

Launder

Choice of timber. The disposition and orientation of the knots and branch stubs on the exterior suggests that the north end of this timber (047 (section B)) lay towards the butt end of the tree. Very little trimming of this had been carried out and in places the bark survived. The absence of a regularly facetted finish, as is found on some Bronze Age woodwork, suggests that the approach to the construction and use of this object was strictly utilitarian. The annual rings were examined at the clearest break and these appeared to be about 2-4 mm apart i.e. of medium growth for alder. A tentative reconstruction of the parent tree is shown in Fig. 83. It seems likely that this would have been found growing next to a stream channel upon the lower slopes, if not along the river bank on the valley floor.

Alder is a soft, relatively easily worked wood, but has little natural rot resistance or resistance to mechanical damage. The clarity of toolmark preservation in this case was for the most part good; no clear signatures (ridges left by gaps in the blade) can be seen, but there are many facets and 'stop marks'(a negative of the blade edge shape), implying that the timber was quickly buried in wet anaerobic deposits before decay could set in. The high level of metal contamination may also have contributed to its survival.

The indicated tool kit. The orientation of the toolmarks indicate that the tool(s) used was hafted as an axe, no trace of marks indicating adze-style hafting were found, as has been the case for some hollowed and sculpted Bronze Age timbers such as elements of the Dover boat (Clark 1997). In practice there are no technical problems hafting flat, flanged or palstave 'axe' heads as adzes if needed. However, such hafting was probably reserved for rather specialized work.

The broad character of the toolmarks can be summarized as follows; short facets and occasional stop marks on the internal sides made by a thin curved blade with angular, mainly oblique cuts (cuts left by the use of an edge tool at a steep angle) in the internal base. Additionally, the branch stubs on the exterior were slightly facetted, and section D at the south end was clearly cut across with an axe and thus may well be the 'felled end'. The three largest axe stop marks, found within sections C were approximately 75 mm wide and convexly curved by about 10 mm as measured from the latex moulds taken. The fineness and width of the stop marks clearly indicates the use of a metal axe. Similarities were noted with the stop marks recorded on the timbers of the Corlea 6 trackway in Ireland dating to about 2268-2250 BC (O'Sullivan 1997), but the Copa Hill examples were about 10 mm wider and very slightly less curved. It may be useful to note here that Late Bronze Age axe marks are generally much smaller than this, for example 35-50 mm wide within the wood from London and Flag Fen (Taylor 1992, 481; Goodburn *forthcoming*). The Copa Hill edge tool used was probably a flat or flanged copper alloy 'axe head' hafted as an axe. This must have been supplemented with wooden wedges and a large mallet for splitting off bigger lumps of waste wood.

The construction of the launder. At the felling site the branches would have been roughly lopped off and the desired length axe-cut (Fig. 83 A &B). Sections of waste were then split off from the top face, probably with some notching and splitting (C), leaving some two thirds of its volume intact. As suggested by the crispness of the toolmarks and lack of ancient decay and borer damage, all of this work would have been done with the wood freshly felled or 'green'. However, it is likely that the launder was allowed to dry for a few weeks after finishing in order to reduce its weight considerably before carriage up the valley side.

Experience suggests that the most likely method of hollowing out the wood would also have been through some method of notch and chop aided by controlled splitting (groove and splinter). The light weight of the metal axes prohibits their use as 'swung wedges', hence a mallet and wooden wedges would have been used for this purpose (Fig. 83 D-F). Towards the end of working the hollow it is clear from the pattern of incuts that the axe(s) were again being used, this time to cut weakening nicks in the base whilst parting blows were used on the sides. The base of the inside would thus have been rather rough, potentially slowing the flow of water. As no tool marks were found corresponding to the smoothing of this surface, it appears most likely that hardwood wedges were used to even the base and break off any remaining loose material (tools of bone or antler would probably have left clear facets). Broadly similar patterns of axe marks have also been recorded in the making of medieval dugout drains (Goodburn & Minkin *in press*).

The second alder launder. This was found in a fragmentary, substantially abraded and decayed condition yet it appeared to be of fairly similar construction to the first. The timber was very knotty, suggesting that it was either a 'top log' or else was cut from a rather heavily branched tree growing in a fairly open location.

General comments on other woodwork (S.Timberlake and D.Goodburn)

Many of the unburnt or partially charred pieces of wood fuel retain some evidence of being cut with a metal axe, either in the shape of small chop marks, side trimming, or cut terminations. The latter may be classified as examples of chisel-shaped, wedge-shaped and 'pencil point' ends, and show some correlation with branch thickness, density and species. For example, simple chisel shaped cuts are more commonly found upon the ends of the relatively narrower and softer hazel stems, whilst more wedge shaped or compound cuts appear in the oak, an analogous situation to that observed within the Mt. Gabriel wood assemblage (O'Brien 1994, 139-140). One piece of hazel (CH96:w10; SEE Fig. 77 B) shows evidence of having been carefully cut around its circumference with a small axe with a rounded blade edge, the rod being only partly cut through then snapped leaving a small projecting tenon or boss in the middle. Scar striations or signature marks produced by some of the sharper axe blades survive upon several small pieces of fuel wood, on some cut rods of hazel (CH95:w4), the point of a small oak stake found on the floor of the mine gallery (CH90:w11;Fig.77 C), as well as on a number of unburnt oak wood chips or shavings (e.g. CH99:w9; Fig. 77 A-E). The latter could have been associated with the lighting of the firesetting hearths.

Figure 82: Drawing of the shape and dimensions of the type of flat axe used and its suggested mounting (ST after D.Goodburn)

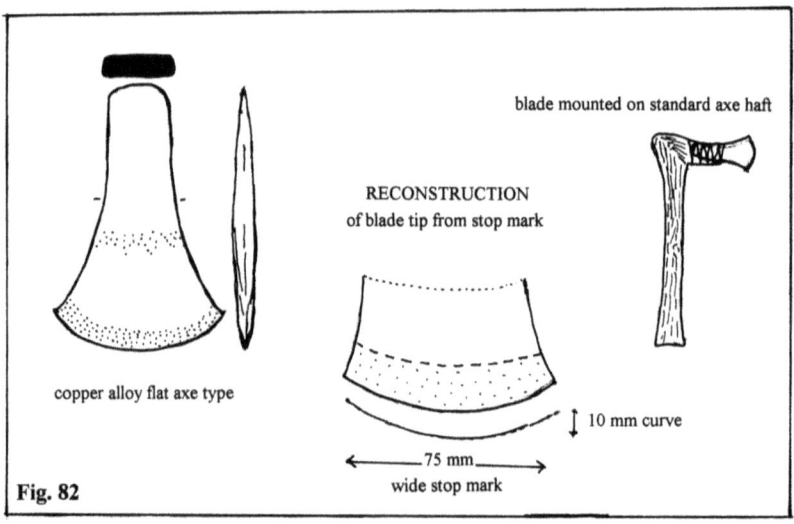

Figure 82b: Detail of axe stop-marks on interior of alder launder (A.Gwilt, National Museum of Wales)

Figure 83: Diagram of parent alder tree and suggested stages in the construction of wooden launder (047). ST after D.Goodburn.

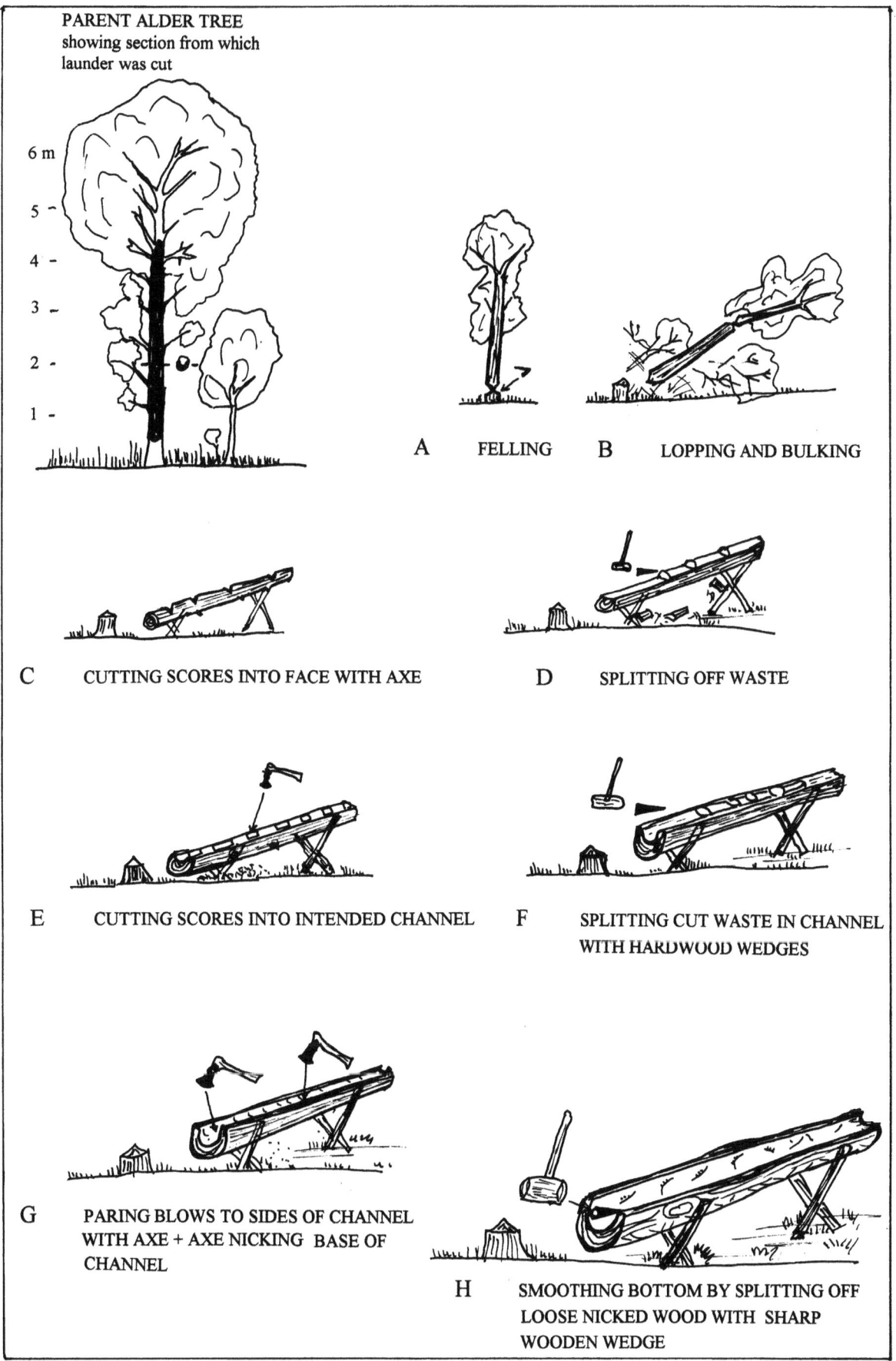

83

CHAPTER 13

ANTLER FINDS (S. Timberlake)

One complete antler pick/hammer (CH95:i5: Fig. 84 A), two large broken tine fragments ((CH96:i2 & CH95:i4; Fig. 84 B), plus several detached flakes or splinters were recovered from amongst the fine-grained mine sediments lying on the floor of the mine entrance. All of these pieces were well preserved and mineralised, one of the tines (B) being stained with copper salts and burnt at one end, but otherwise showing little evidence of use. A few rather decayed and fragmentary remains of broken antler were also found within the basal layer of the Central Tip (Timberlake & Switsur 1988) and also in the middle of the earliest Lateral Tip located outside the opencast. All were of red deer (*Cervus elaphus*).

An examination of the pick/hammer (A) suggests a long history of use and re-use, greater still perhaps than many comparable examples recorded from flint mines and other sites of prehistoric quarrying (Sieveking 1979; Holgate 1991). This implement had been selected and made from a fairly substantial piece of antler measuring some 30 cm from the tip of the crown, with the shaft at the handle end cleanly broken, and with the trez tine sawn or cut off at the join (the stub here showing evidence of a compound axe cut) for ease of holding. The base of the broken narrower bez tine had been left projecting several centimetres, perhaps also to allow a better hand grip and for additional protection. Some evidence of wear was visible on this 'handle', the pattern of this suggesting right-handed use. Wear at the pick end was extremely heavy, with the brow tine worn down almost to the level of the shaft and further evidence for wear on the crown, the rim of which was smooth completely rounded on one side. Alongside damage to the upper half of the shaft on the tine side, the pattern of wear suggests that the implement was also used as a hammer or mallet (on the flattest side of the shafthead as well as upon the tip and front of the crown), probably subsequent to its use as a pick. The tool was discarded at the point at which it was considered to be no longer serviceable as either. As with stone tools, the type and degree of usage suggests that the approach to these implements was totally utilitarian. For instance, the reconstruction and experimental use of an antler tool of very similar appearance has shown that less than a third the same amount of wear to the brow tine might be expected (when used only as a pick) following the mining of about 5 tons of fireset rock (Timberlake 1990a, 54). Within some Neolithic flint mines there is less evidence for re-use, and more for specialized tools, for example antler picks *as well as* antler hammers, bone points and shovels (Holgate 1991).

Although picks of hartebeest horn were first used some 35,000 years ago to extract chert from underground galleries within the mines of the Nile Valley (Vermeersch & Paulissen 1989, 35-36), the widespread use of picks made of red deer antler is well attested in the Neolithic flint mines of Europe, as suggested by the numerous finds from Grimes Graves, Hambledon Hill, Blackpatch and other sites in Southern England (Holgate 1991), and more graphically still by the discovery of a crushed skeleton of a miner found holding a single-handed pick in a mine near Oubourg in Belgium (Bromehead 1950). Unfortunately, evidence for their use in metal mines has been harder to find. Part of the explanation for this may lie with their preservation (this appears to have been much better in chalk than in the more acidic mine waste often associated with the extraction of metal (sulphide) ores), although it should also be noted that considerably fewer Bronze Age metal mines have been excavated. Nevertheless, antler implements were thought to have been the main mining tools within the Eneolithic (5th millenium BC) copper mines at Aibunar in Bulgaria (Cernych 1978), whilst others were found at Rudna Glava, Yugoslavia (Jovanovic 1979), and in the Beaker period copper mine at El Aramo in Asturias (Hedges et al. 1990 -see O'Brien1994). Tool marks interpreted as being made by antler picks or wedges have also been described from Siphnos, Greece (Wagner et al. 1980). Within Britain, the substantial assemblage of bone tools from the Great Orme has produced very little in the way of utilised antler, no more than a couple of broken tines amongst thousands of pieces of bone (the latter presumably better suited to scraping the ore from the soft dolomite (*pers com*. A.Lewis)); although a 'broken piece of stag's horn' was recovered from the prehistoric mine at Nantyreira, mid-Wales in 1859 (Timberlake 1988,14), and more recently within the workings of the prehistoric-Roman copper mine at Ogof Llanymynech (Adams 1992). However, neither of these latter finds have been dated. Evidence for the use of an antler tine as a replaceable pick end (slotted into a either a wooden or antler shaft) comes from Ecton Copper Mine in Staffordshire, where a fragment recently discovered amongst rubble within shallow workings underground has been dated to the first half of the 2nd millenium BC (Barnatt & Thomas 1998). The latter find suggests a utilitarian solution to the sort of extreme wear witnessed on the Copa Hill antler tool, even though similar examples of picks with replaceable points are also known from the Neolithic (Sieveking 1979, 6).

Notwithstanding the incomplete survival of antler artefacts on Copa Hill, yet armed with some clues as to their function and probable life expectancy, it is possible to estimate the probable numbers of tools involved. Assuming that only a proportion were used in mining/firesetting operations, and that each pick could have assisted in the removal of between 15 and 25 tons of rock, it is conceivable then that between 100 and 300 antlers may have been brought up to site. Over several hundred years of intermittent working this is not a large amount, particularly if this involved the collection of shed rather than butchered antler. Unfortunately, very little is known of prehistoric population of native red deer within the Welsh uplands, although their decline in numbers from the early medieval period onwards was almost certainly linked to the loss of woodland cover, in particular to the creation of the great Cistercian sheep walks (Condry 1994).

Figure 84: Antler artefacts from Entrance A (D8): (**A**) pick/hammer [CH95:i5 (077)]; (**B**) broken end tine (pick?) [CH95:i4]. Drawings B.Craddock, 1995 & 1996.

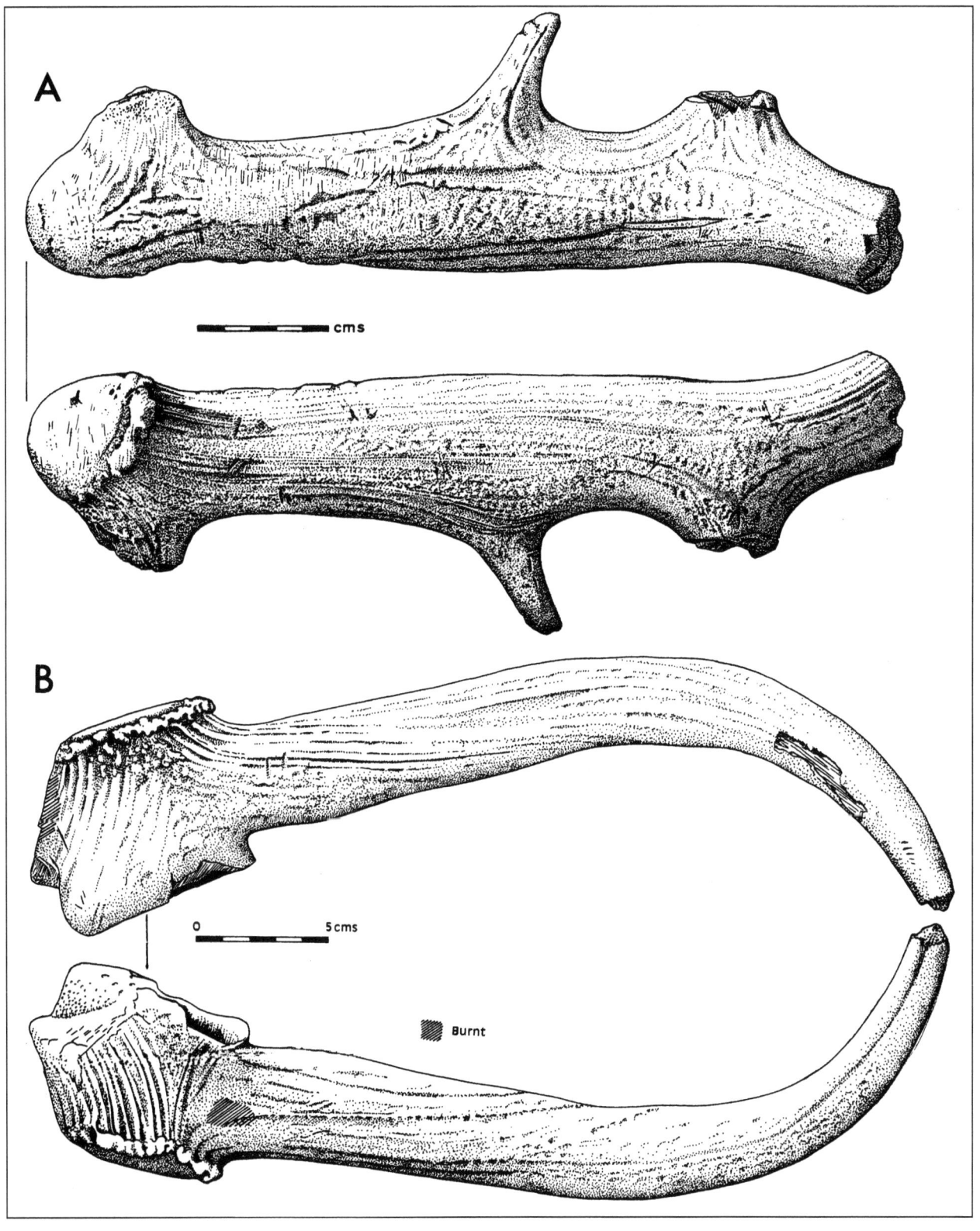

CHAPTER 14

HAMMER-STONES (Simon Timberlake & Brenda Craddock)

Cobblestone hammers or mining mauls, plus the broken and commonly re-used fragments of these, comprise the bulk of the mining implements used at Cwmystwyth. The study of these has provided us with the greatest amount of information about prehistoric mining technology, as well as of the processes and the location of mining, ore separation and dressing. Analysis of cobble selection criteria, tool wear, use, and modification (for hafting) is currently underway, the initial results of which are presented here in brief.

Since 1986 a total of **1203** unbroken or partial remains of hammer-stones (the latter including many discarded 'core' tools, large flakes or spalls) have been excavated or else recorded *in situ* upon the surface of the prehistoric tips. In 1989 the position of some 250 tools were plotted upon a plan of the external spoil tips, whilst their weight and length/width measurements were compared and a provisional typology created based upon a crude morphological grouping of similar sized, shaped and used implements (Timberlake 1990a). Later in 1995 Gale carried out a series of morphometric analyses on the above assemblage, as well as upon another 90 hammers deposited within the collections of the British Museum and the National Museum of Wales, most of which had been collected during the 1986 and 1989 excavations (along with those from O.Davies' excavations of 1935). This work included the results of number of statistical analyses of cobble shape, sphericity, fracture, wear patterns and modification type, but very little on the geological parameters such as petrology, grain size, silica content and hardness of the cobble rock types. In another study, examples taken from Cwmystwyth and other sites in England and Wales were used to help establish a broader classification of the prehistoric mining hammer, for example a typology based almost wholly upon the degree of artefaction or subsequent modification of these tools (in the form of pecking, notching or grooving), the latter presumably for the purposes of hafting or attaching a handle (Types 1-6: Pickin 1988;1990). Key elements within both of these latter two typologies have since been included within a new record sheet designed for recording and classifying mining hammer-stones from an ever-growing number of British sites (similar criteria have also been used for classifying stone tools at Mt.Gabriel (O'Brien 1994, 124-127).These record sheets have recently been trial tested as part of a re-classification of the collection of grooved and un-grooved mauls from Alderley Edge present within the historic Manchester Museum collections, as well as from recent excavations at Engine Vein (Timberlake *forthcoming*). From Copa Hill, all 953 implements recovered from the 1986-1999 archaeological excavations have now been examined in an identical way, the initial results of which are enclosed.

Recording hammer-stones

Each of the Copa Hill tools was individually assessed against 41 easily measurable parameters. A simple description of these criteria, most of which could be measured fairly quickly and accurately with the aid of weighing scales, calipers, ruler, hand lens, a simple hardness kit, and a prepared set of visual standards is presented in Appendix 1. Linked to these record sheets there is a comprehensive finds archive which includes photographs, drawn profiles, and a number of finished drawings and notes (B.Craddock). The latter are in the process of being deposited with the Department of Archaeology, National Museum and Galleries of Wales, Cardiff.

The selection and sourcing of cobbles

Previous visual examinations of cobbles used as hammer-stones on Copa Hill had already suggested to us that there was some selection practised in the choice of suitable material (Pickin & Timberlake 1988). Moreover, an objectively based metrical study of cobbles carried out at several possible source sites including the bed load of the River Ystwyth opposite the mine, and a glacial till at Lan Fawr some 1.5 km to the north of here, seems to confirm the small percentage of suitable cobbles available locally (Gale 1995). Indeed, the lack of glacial abrasion and angularity of the till cobbles would appear to exclude local moraine as a direct source. Moreover, the roundness indices of some 36% of the Copa Hill hammer cobbles clearly exceeded those of pebbles from the River Ystwyth, suggesting an additional or altogether different source for some of this material (Gale 1995, 255). A rather similar result was obtained by ourselves following the sampling of a much larger number of hammers recovered from the excavations; some 33% proved to be of maximum roundness (9) on a visual roundness scale of 1 to 9 (Krumbein 1941) [SEE Appendix 1, p.127], thus re-affirming the suggestion that these cobbles had been selected, at least in part, on the basis of roundness. However, this conclusion was still based upon the assumption that the source(s) of these implements were river cobbles. The analysis of the cobble surface has been much more revealing.

The high degree of polish present upon the surface of many of these had already been used as a basis for suggesting that at least some were beach pebbles brought from the coast (Timberlake 1990), a fact which had not escaped the attention of Lewis Morris some 250 years earlier, when describing the stone hammers at Twll y mwyn (Bick and Wyn Davies 1994). In 1996 a study was carried out to make visual comparison between a selection of Copa Hill cobble hammer-stones and typical alluvial cobbles observed at various points along the Ystwyth floodplain between Cwmystwyth and the coast some 25 km away at Tan y bwlch beach, Aberystwyth (Jenkins & Timberlake 1997). The only good correlation found was with the samples of beach cobbles, even though many of the latter appeared to be flatter and less cylindrical/ovoid in shape than the cobble stones found used on Copa Hill. However, it was interesting to find that from amongst the current excavated sample of 723 hammer-stones which had been successfully measured for smoothness, some 234 were classified as having a polish recognisably equivalent to beach pebbles (scale 1), with another 339 being highly smooth (scale 2) - suggesting that up to 79% of these cobbles *may* have been brought here from the coast. Some also show evidence of the 'chatter marks' or chinks typical of beach pebble attrition. At least 5% of the remaining cobbles, however, do show the distinctive characteristics of river pebbles. Analysis of cobble shape does seem to indicate a high incidence of pebbles with flat surfaces (>40%), although a significant number of rectangular, cylindrical, and ovoid forms may suggest to us that these were the sorts of cobbles preferentially selected, since such designs are proportionately rarer within the littoral bed-load.

Dimension (length/width/depth) as well as weight of cobbles has been used as a common means for empirical comparison between assemblages of hammer-tones from different mining sites (Gale 1995; O'Brien 1994). In this particular respect the measurement of 900 of the most intact cobbles from Copa Hill is quite significant in that it shows the median weight of selected cobbles to be about 2 - 2.25 kg (with the largest being about 8 kg and the smallest weighing as little as 250 g), a figure considerably greater than the average of 0.75-1 kg formerly recorded from amongst a sample of 195 hammers (some of which may have been considerably more broken examples) found lying upon the surface of the tips in 1989 (Timberlake 1990 b). Much lower average weights have been reported for undamaged hammers from Alderley Edge (1-1.5 kg), Mt.Gabriel (1-1.5 kg), Parys Mountain and Nantyreira, although at least three different sizes of hammer, including some weighing in excess of 14 kg, have been recorded from the Great Orme (Gale 1995; O'Brien 1994). Factors such as the average size and weight of the source cobbles, the breakage rate of these at the mine, and the degree of selection carried out for the specific tool requirements of the site may all have determined in some way the actual sizes and weights of cobbles chosen. Not surprisingly, there appears to be a much larger and more continuous variation in the size of hammers found upon Copa Hill, the majority of which are cobbles between 15-25 cm long (mean length 20 cm) and 8-13 cm wide (SEE Fig.86) - greater than the average length of cobbles used at Parys Mountain and Mt.Gabriel (14-16 cm), yet of similar width and thickness. It is tempting therefore to speculate whether larger hammers correlate with predominantly opencast workings. However, the workings on Parys Mountain were probably always in part at least, opencast, and in hard rocks (Timberlake 1988; Jenkins 1995). It seems more reasonable to suggest that such differences in size may have resulted from similar limitations present within the supply of available cobbles.

Some preferential selection of fine grained rocks also seems to have been undertaken by the Copa Hill miners, as suggested by the frequency of cobbles of dark grey greywacke. Certainly, the composition of rock types present in a sample of 784 hammer-stones which could be examined petrologically seem to be skewed this way when compared with the 'natural distribution' of local sandstone/greywacke boulders, although it is rather more difficult to determine in this case whether or not this reflects the composition of a more distant source such as a beach deposit. All of these cobble rock types were identified in hand specimen. Some 42% were composed of fine-medium grained quartzitic greywackes and sandstones (series Cb), whilst another 19% were hard quartzitic grits (D), and a further 8% were of 'local' orthoquartzites (E). Fissile flagstones or rocks with prominent bedding laminae, or those with cleavage, obvious joints or quartz veinlets were much less common, perhaps rejected because of their shorter life expectancy and risk of premature fracture. Nevertheless, the sizeable presence here of many such cobbles subsequently rendered useless following premature planar fracture along joints or laminations after a short interval of use (>10%) testifies to the lack of any really vigilant degree of selection, or perhaps even to an understandable inability to detect flawed stones. The relatively limited life expectancy of such hammer-ones within this hard quartz-rich mine rock environment may itself have been the reason for a rather slack selection criteria, or perhaps the miners did not always choose this material themselves. The choice of fine-grained quartzitic sandstones must have helped to avoid some of the above problems and increase longevity of use and this correlates well with the overall high incidence of conchoidal fracture (63%) witnessed.

Storm beaches may well have made superior collecting sites for hammer-stones, localities at which there was the choice to select right-sized/shaped implements from amongst the many thousands of cobbles previously sorted under turbulent conditions, resulting in the preservation of the most competent examples. It is perhaps rather surprising then that we do not find more examples of igneous or meta-sediment cobbles on Copa Hill, given that the nearest coastal sources at Tan y bwlch and Morfa Bychan contain upwards of 5% exotic pebbles (mostly of re-worked glacial erratics). Nevertheless a number of hammer-stones of igneous or non-local quartzite rocks have been recovered from the mine (<1% of sample), the only feasible source of which could have been the coastline south of Aberystywth where the nearest deposits of non-local moraine left by the Irish Sea ice carrying igneous material from North Wales, the North of England and Scotland are to be found (Williams 1927). The petrology of one of such tools (CH89:h21) was examined in thin-section and this proved to be a garnet and epidote bearing quartz porphyry of a type typically found amongst the rocks of the Borrowdale Volcanic series in Central Cumbria (Humphries 1989). A recent beach survey (SEE Timberlake & Jenkins 1997) has shown that the average size of most of the far-travelled erratic pebbles found lying upon the storm beaches was substantially smaller than the optimum chosen for use at the mine, providing a possible explanation for their rarity as hammers. Coincidentally, greywacke or quartzitic sandstones appear to amongst the commonest rock types found used as tools within the prehistoric metal mines of Britain. Even at Alderley Edge, where a wide range of Lake District erratic cobbles were available from the Local Drift, the heavier igneous rocks do not appear to have been chosen preferentially over the greywackes (Timberlake *forthcoming*).

Tool types, tool use and re-use

Functional and experimental studies on the hammer-stones and discarded stone tool fragments from Cwmystwyth have been carried out since 1989 (Timberlake 1990a, 1990b; Gale 1994; B.Craddock 1994b; Craddock & Craddock 1996, 1997). These studies are continuing, and a summary of their conclusions are presented here in brief.

Gale recognized at least 4 tool types on Copa Hill, hammers, modified end-hammers, hammers and anvils/mortars, and flake tools, and in addition noticed a much greater level of re-use and of broken discarded tool material here than he had observed at his other study sites at the Great Orme, Alderley Edge, Parys Mountain, a fact which he put down to the relative isolation of this site, the type of cobbles available, and the hardness of the quartz-rich rocks worked (Gale 1994). In fact, the present study suggests that at least **41%** of the hammers found within the mine appear to have been re-used following initial breakage, a figure which might well turn out to be higher, but of course not always easy to ascertain when we are dealing with what are often quite incomplete fragments. Of equal interest is the division between identifiable tool functions and their frequency of use within the implement assemblage (this breakdown therefore does not refer to actual numbers of implements). For example, amongst a sample of 892 implements at least 86% showed evidence for substantial use at either end as *mining hammers or picks* (for rock breakage), whilst some 8% of these had been used/also used as *hand-held pounding/crushing implements* (most probably for ore processing or careful vein extraction work), 10% as *mallets* (hand-held for use with chisels, wedges or other implements), 7% as *chisels, wedges* or other sort of flake tool, 17% as *crushing anvils* (for ore processing), and a very small number (< 1%) as specialised *pecking tools* (perhaps for the notching and modification of hammer-stones). Only 9% of all tools showed indisputable evidence of modification for the purposes of being hafted with a flexible handle (75 examples were edge notched around the mid-rift of the cobble, whilst 4 were more obviously semi-grooved). In contrast, O'Brien (1990) claimed that some 30% of the Mt.Gabriel tools were bi-laterally notched for hafting. It is interesting therefore that experimentation has since demonstrated that such modification of the cobble surface is not always necessary for this to be achieved (Pickin & Timberlake 1988; Craddock & Craddock 1996).

A more detailed alpha-numerical classification of hammer-stone function/modification for Copa Hill (similar to that carried out for Alderley Edge) is presented below within a considerably summarised and simplified form. In accordance with Pickin's classification of mining hammers (Pickin 1988, 1990), these would now be classified as follows e.g. 1AA (unmodified), 2AA (notched), 3 AA (semi-grooved) etc. The number of hammer-stones recorded (799) reflects those sufficiently intact enough to be classified.

Figure 85: Typical association of hammer-stone [CH90:h14] with charcoal and wood debris amongst re-deposited mine spoil (D3 031) inside of opencast. Tool is 18 cm long Photo ST 1990

Functional use/ wear/ re-use categories assigned to tools	unmodified (class 1)	notched for hafting (class 2)	semi-grooved (class 3)	Total
A (single end hammer)	431	-	-	431
AA (double end hammer)	70	30	2	102
C (broken end as hand crusher)	10	-	-	10
D (flake re-use as chisel etc.)	17	3	-	20
E (anvil stone only)	2	-	-	2
F (pecking stone for notching?)	2	-	-	2
AB (hammer + side use as mallet)	30	5	-	35
AC (hammer + hand held crusher)	24	7	-	31
AD (broke hammer re-use as chisel)	25	10	1	36
AE (hammer + use of side as anvil)	65	8	-	73
ABE (hammer + mallet and anvil)	21	6	-	27
ACE (hammer + hand crusher/anvil)	18	3	-	21
ADE (hammer + chisel + anvil)	4	2	-	6
ABD (hammer + mallet + chisel)	-	1	-	1
ABCD/E (compound tool- re-use)	2	-	-	2
TOTAL NUMBER OF HAMMERS	721	75	3	799

Table 3 : Alpha-numerical classification for hammer-stone function/ modification for Copa Hill (ST 2001)

Figure 86: Length/width plot for Copa Hill hammer-stones

Figure 87: Weight (kg) frequencies for Copa Hill hammer-stones

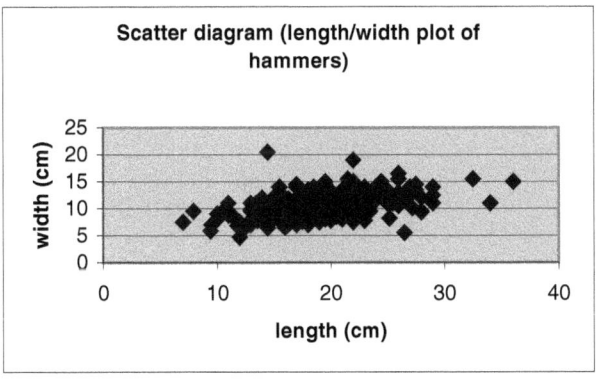

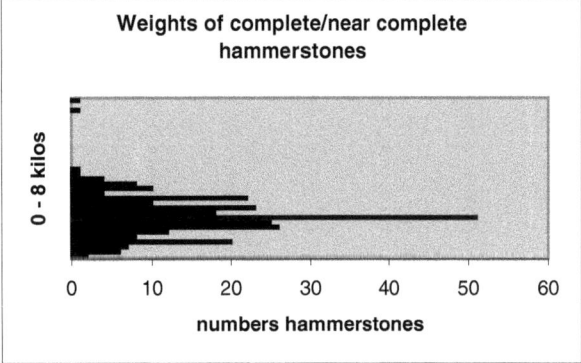

Figure 88: Chart of cobble shape frequencies for Copa Hill

Figure 89: Cobble smoothness frequencies for selected cobbles

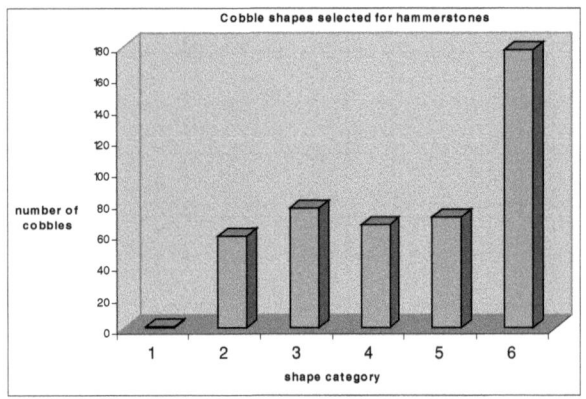
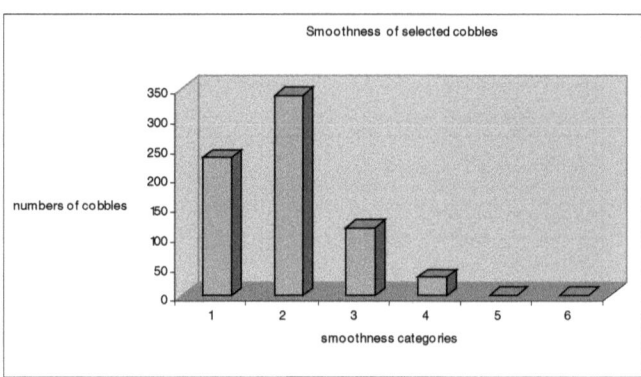

Figure 90: Suggested tool function(s) within analysed assemblage of used cobbles

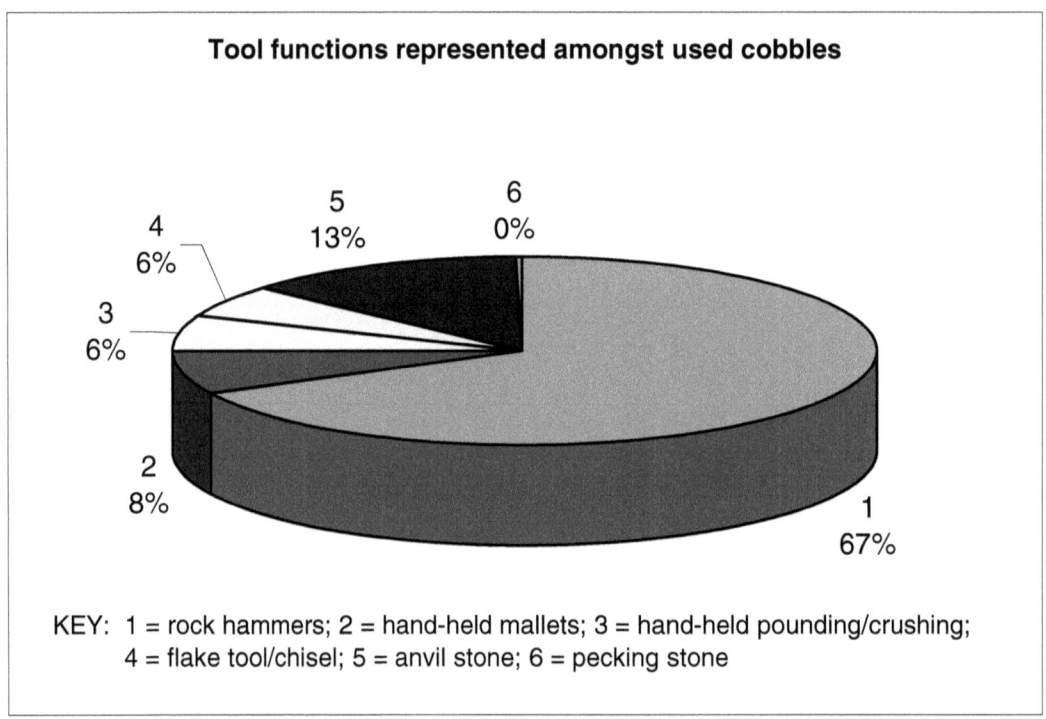

Analysis of the distribution of implements showed almost no association between tool types and particular areas of the opencast mine, although it should be recognized that less than ten of these were found in locations where they were clearly *in situ* (such as in the place where they were last worked), most forming part of a mobile entity which had been used, re-used, discarded, moved around in spoil and then finally tipped wherever convenient. This loss of original context is also a relevant issue when addressing the suggestion that copper ore may once have been crushed *in situ* upon the surface of the tips outside the mine (Mighall et al. 2000), since the data from the 1989 tip survey failed to provide any evidence for the larger number of anvil tools which should be expected. In fact, similar numbers of these appear to have been discarded within spoil backfilled or slumped into the interior of the mine. There also appeared to be no clear correlation between those hammers re-used as anvil stones and the incidence of coarse grain rock types (such as one might have expected of more specialized tool kits containing specific mortar stones and querns), although some 66% of all the hammers used or re-used as anvils were composed of quartzitic sandstone or quartzite (more than 64% of these had a hardness > 5.5), whilst at least 45% of these cobbles were flat sided. However, less than half of this use was primary and associated with flat (Type 6) or rectangular shaped (Type 5) pebbles (Fig. 96 B), much of it instead representing an opportunistic re-use of the flat surfaces provided by the longitudinal breakage of hammer-stones, or else of the scars left behind by the detachment of large flakes (Fig.96 C). All this suggests a fairly utilitarian approach to using and re-using available cobbles and broken material, with relatively little fixation on maintaining quotas of specialized tools. Nevertheless, there are individual examples which appear to have experienced a long currency of use at the mine (as indicated by the degree of wear, particularly faceting of hammered surfaces), possibly because the cobbles themselves had proved unusually competent, but perhaps also because their shape, size, weight and appearance had helped to make them favoured implements. Most of the latter appear to have been hand-held.

A few of those tools with specialized functions may be associated with the several *in situ* environments within the mine. For example, a small cache of hammers including a hafted chisel-pick (Fig.94 A) were recovered from a tight vein working within the rear of the perched mining gallery (D3), whilst several small waste flakes were found detached as a result of mining/crushing activities carried out upon the floors of the rock platforms. A single flake tool notched upon its sides (Fig.94 B), perhaps for holding with a handle whilst being struck with a mallet, was found jammed between the sides of a small lead vein (Fissure 2) within the floor of the entrance cutting (Fig 99). This remains the only example of a stone implement found abandoned whilst in use, although it should be noted that several bone tools have also been found *in situ* within dolomite fissures underground on the Great Orme (Dutton & Fasham 1994; Davies, G.C. 1996).

The relatively minimalist approach to haft modification witnessed upon the surfaces of these cobble tools could just be another response to their much shorter life expectancy (when compared for example with the grooved implements of Alderley Edge), yet it remains an important task to understand exactly how these implements were used. Experimentation has shown that well rounded rather than angular cobbles provide a much better grip for the twisted and bound withy handles. Furthermore, a comparison of hammer-stone wear and shape has shown a good correlation between ovoid/pear-shaped cobbles and the use of their broadest ends as unmodified hammer-stones (>70% of examples). Used this way the cobble would not require any sort of modification - instead it would be wedged further back into the single or double loop of twisted withy with each successive blow, thereby tightening the grip of the haft (SEE B.Craddock 1994 b). Subsequent use of the narrow end as a pick or hammer would also be possible (SEE Fig.92 B), but this would require notching or faceting of the edges (sides) of the cobble and perhaps also the use of a stone or wooden wedge to help tighten the haft. The most pronounced notching or partial grooving is thus to be found within those examples of hammer-stones (commonly cylindrical or rectangular cobble forms) heavily used at both ends (Figs.93 C,D&E). The typical response to extensive breakage and loss of original surface around the mid-rift of the cobble was to re-notch the edges and perhaps also the flatter flaked side of the hammer (e.g. Fig.93 C), although little more than a scratching or rough abrasion of these areas was sometimes necessary to round off sharp edges and enhance the grip of the haft. Thus it has also been suggested that some of these lateral areas of scratching were a means to secure the haft-tightening wedges, or else were the marks produced from hammering these in (Craddock & Craddock 1996, 7).

Flakes or reduced hammer-stone cores which have been re-used as chisel or wedge tools show obvious signs of subsequent blunting and scratching upon their points, as well as the impact marks from hammering, and occasionally also the modification notches for a handle. Hand-held tools, on the other hand, are commonly chosen from amongst the broken tips or ends of hammers (Type C: e.g. Fig.95 A), although very occasionally palm-sized pebbles including double or triple pointed forms are found (Fig.95 B), many of which appear to have been used on most available faces as well as along the corner edges around the circumference, presumably as implements to cob small fragments of ore (rather similar disk-shaped pebble hammers are illustrated from both the Great Orme and Mt.Gabriel mines (O'Brien 1994; Dutton & Fasham 1994)). These smaller crushing implements (usually < 0.5 kg) as well as the larger pounders were probably used in association with the anvils. The latter were sometimes used on both sides, or again upon the same face, the area of crushing being typically accompanied by a mineral (iron oxide) stain, but with indications of abrasion varying from a faint scratching to indentations of up to 5 cm wide and several millimetres deep (e.g. Fig. 96 A). A number of the heavier crushing tools also appeared to have areas of mineral staining as well as the development of facets upon their ends. These may have resulted from the repeated pounding action necessary for the extraction of good ore from vein rock and the comminution of the sulphides.

It seems likely that broken hammer-stones or collections of larger flakes may well have been kept in caches on site, in order to maintain of pool of readily re-usable material from which 'a tool for the job at hand' could be readily selected (info. B.Craddock).

Discussion - the use of stone hammers in mining

The earliest use of hammer-stones in mining stretches back some 35000 years to the excavations of underground galleries for the extraction of chert in the Nile Valley (Veermeersch & Paulissen 1989), although the use of similar tools within the hematite pigment mines of Bonsu, Swaziland 28000 years ago suggests an association with metallic ores very nearly as ancient (Bick 1995 HMS News). Hammer-stones were commonplace as implements within the Neolithic flint mines of Southern England and mainland Europe (Bromehead, 1954; Holgate, 1991) as well as in the quarries and axe-factory sites of Cumbria (Claris & Quatermaine, 1989), yet the appearance of recognisable mining 'mauls' or hammer-stones (the use of elongate and commonly grooved examples of cobbles) seems to be associated with the earliest mining for copper ores and the smelting of metal. Whilst not the earliest copper mines exploited, some of the more familiar Chalcolithic sites with which such hammer-stones are associated are to be found within the Balkans and the Near East. These include Aibunar, Bulgaria (4700 BC), Rudna Glava, Yugoslavia (4000 BC), Wadi Feinan, Jordan and Veshnoveh in Iran (3200-3000 BC) (Cernych 1978; Jovanovich 1979; Hauptmann 1990 ; Holzer et al. 1971). Early dates have also been obtained from several mines within SW Europe, notably from El Aramo in Asturias, Spain (4000 BC), Libiola near Genoa, Italy (3350 BC), and Cabrieres near Clermont Herault, France (3000 - 2500 BC) (Hedges et al. 1990; Maggi & Del Lucchese 1988; Ambert et al. 1983), evidence which suggests a fairly rapid diffusion of mining and metalworking technology westwards, or possibly its quite independent initiation within different areas (Craddock 1995, 144-145). Grooved hammer-stones also appear in profusion at Ross Island, Killarney, in Ireland where miners were both extracting and smelting copper sulphides and fahlerz ores by 2500 BC (O,Brien 1996). Within the British Isles as a whole there appears to be no convincing evidence for their use in metal mines later than the end of the Early Bronze Age (c.1500 BC), although in Central and Southern Europe several isolated examples of their use continue into the Late Bronze Age, for instance at the mines within the Shwarz/Brixlegg area of North Tyrol from which both grooved and notched mining hammers as well as dressing stones have now been reported (Reiser & Schrattenthaler 1995), as well as SW Spain where the trench mine complex at Chinflon has also returned Late Bronze Age dates (Andrews 1994; Rothenberg, R. & Blanco Freijeiro, A., 1980). Metal tools (bronze picks or chisels) supercede stone mining tools at Thorikos in Greece, although here it is believed that the use of stone hammers continued into the Iron Age as crushing tools (Mussche et al. 1990). Similar implements were also used during the Late Bronze Age for ore processing at the Cypriot copper mines (Knapp 1999), and there are claims for their continued use within mines of the Punic and Roman periods in Spain (Rothenberg & Blanco Freijero, 1981). Medieval and later use of stone hammers as mining implements is reported from sub-Saharan Africa, for example within workings for malachite and azurite dating from the 13th century AD at Harmony Mine in the Transvaal, as well as at nearby Phalaborwa (770 - 1000 AD), whilst in Zimbabwe these tools were used for crushing gold and copper ores right up until the eighteenth century (Herbert 1984). Similar stones were used for crushing gold ores in the fourteenth century at Tembelini in Mali (Ford 1998), whilst the occasional use of hand-held hammers or pestles continues even today as part of small-scale village processing of gold-bearing quartz rock in Uganda (Worthington & Craddock 1996). Nevertheless, despite assertions by Peake (1937) that 'stone hammers....were used in Britain from the Bronze Age to Iron Age, throughout Roman times and well beyond' (with similar claims made by Warrington 1981 (at Alderley Edge) and Briggs 1988), there seems to be no evidence for their use during the historic period, either within this country or in any of the known European mining fields.

Cobble stone mining hammers (or 'hammer-stones') from prehistoric metal mines within the British Isles have been moderately well investigated, yet the results of this work are not widely known. Furthermore, the subject still suffers from the lack of any standard terminology or definitive descriptions for these tools as a means by which they may be recognised when found out of context from accompanying dateable mine spoil, a problem recently highlighted by the discovery of similar looking implements either as stray finds or in museum collections and provenanced to the metalliferous regions of Cornwall, Cumbria, Snowdonia and Scotland where Bronze Age mines have yet to be recognised. Nevertheless, the study of the Copa Hill assemblage has provided some new insights into the procurement and use of these utilitarian tool sets, a more complete study of which is now underway.

Figure 91: Selection of experimentally hafted cobbles (as mining hammers) using twisted willow (centre), hazel sticks and rawhide, and clematis rope. From the Early Mining Workshop at Plas Tan y Bwlch, Snowdonia in November 1989. Photo B.Craddock.

PLATE FIGURES 92 - 96 - **STONE TOOLS FROM COPA HILL** (pages 94-98)

Figure 92: End-hammers (little modified)

(A) Single end-hammer *A* (unmodified). Cobble shape 2 [CH89:h60 (D2 015)]
(B) Double end-hammer *2AA* (with only minor use of narrow end). Cobble shape 2. Notched for secondary hammer use at broad end [Chex:h205 (D7 013)]
(C) Double end-hammer (little modified). Cobble shape 1 & 2 [CH90:h10]
(D) Single end-hammer (unmodified). Short duration of use (fractured). Cobble shape 5 & 3 [CH89:h39 (C3 0030]
(E) Double end-hammer (little modified) showing heavy and long duration of use. Cobble shape 4 & 5 [CH86:h79 (D1a)]
(F) Double end-hammer (well used). With evidence for re-use after major fracture. Cobble shape 5 [Chex:h234]

Figure 93: End-hammers, picks etc (notched and semi-grooved)

(A) Double end-hammer (probably also used as mallet and anvil on flat surface *(2AACE)*). Prominent bilateral notching [CH89:h5]
(B) Double end-hammer and pick *(2AAD)* with deep notch on one side [CH89:h9]
(C) Double end-hammer/pick with evidence for primary modification (bilateral edge notching) then subsequent grooving across one flake surface following fracture *(3AA)* [CH89:h18]
(D) Double end-hammer with two phases of modification (notching) relating to use of implement at different ends [CH89:h15]. Cobble shape 4.
(E) Grooved hammer/ pick (3AAD) [CH89:h11]
(F) Double end-hammer showing evidence for extended use. Notched/ semi-grooved (3AA). Cobble shape 1 & 3 [CH99:h421]

Figure 94: Flake tools (chisels etc.), waste flakes and spalls

(A) Heavily reduced core of double-end hammer re-used as a small chisel. Bilaterally notched for hafting *(2AAD)*. From mine gallery [CH90:h15 (D3 020)]
(B) A well re-used flake (split hammer) as chisel/prising tool *(2AD)*. Found embedded in galena vein [CH94:h212 (D7 (A) 0520]
(C) Flake re-used as chisel, and possibly as anvil. Notched for hafting (on flake and cobble surfaces) [Chex:h213 (CH95 D7)]
(D) Thin flake used as chisel. Notched on one edge [CH89:h52 (C4 003)]
(E) Oval-shaped flake off of a cobble (hammer) re-used around part of circumference as either a chisel, scraper or prising tool [Chex:h66]
(F) Waste flake (spall) from the tip of a well-used hammer-stone [CH93:h10]
(G) Longitudinal flake or splinter resulting from the bilateral fracture of a hammer-stone [CH95:h5]
(H) Irregular-shaped spall or splinter from the tip of a hammer [CH99:h1]
(I) The detached tip of a narrow-ended hammer-stone [CH90:h6]

Figure 95: Hand crushing/ pecking tools

(A) Broken end of a cylindrical hammer-stone (cobble shape 4) re-used as crushing implement on one end *(C)*. Note indentation in centre of flattened pounding surface caused by wear resulting from the crushing of small cobbed material [CH96:h1]
(B) Broken tip of hammer used as crushing implement. Three areas (tip and broken corners) used. Possibly sides of cobble also, but more likely that it was re-notched for hafting *(2AC?)* [CH89:h41 (D4 022)]
(C) Hand-held crushing implement. Possibly primary use, but conceivably was formerly used as a hammer-stone. Note how most edges of the rectangular cobble have been used for light pounding/crushing work [Chex:h208]
(D) A well-used hand-held crushing implement/mallet. Both sides and ends show evidence for extensive use and re-use [CH89:h31]
(E) Neat hand-held crushing implement. Very probably used also upon each of its sides as a mallet, perhaps with small stone chisels [CH91:h8 (D3 030)]
(F) Pecking-stone made of vein quartz. Possibly used for crushing, but most probably for pecking notches/grooves within the sides of hammer-stones *(F)* [CH96:h2]
(G) Very well-used crushing stone/mallet [CH89:h81 (D4 031/032)]

Figure 96: Anvil stones, 'saddle-quern type' mortar, and miscellaneous (stone lid)

(A) Double end-hammer re-used as an anvil stone upon flattest cobble surface (AAE) [(C4 006)]
(B) End-hammer (of cobble shape 5 & 6) re-used on both upper and lower flattened surfaces as an anvil stone [CH90 unstratified from tips]
(C) Fractured and discarded end-hammer re-used on concave flake surface as an anvil stone [CH89:h7]
(D) 'Saddle-quern type' crushing stone or mortar. Large sandstone slab extensively worn and smoothed (on both upper and lower surfaces) as a result of grinding action, probably using a rubbing stone [CH86 A 2/1 no.117]. Deposited within collections at the National Museum of Wales (1986)
(E) Stone 'lid' of chipped and roughly-shaped fissile sandstone found below turf [(CH89 D2 001/002)]. Donated to British Museum.

Figure 92: STONE TOOLS - End-hammers (little modified)

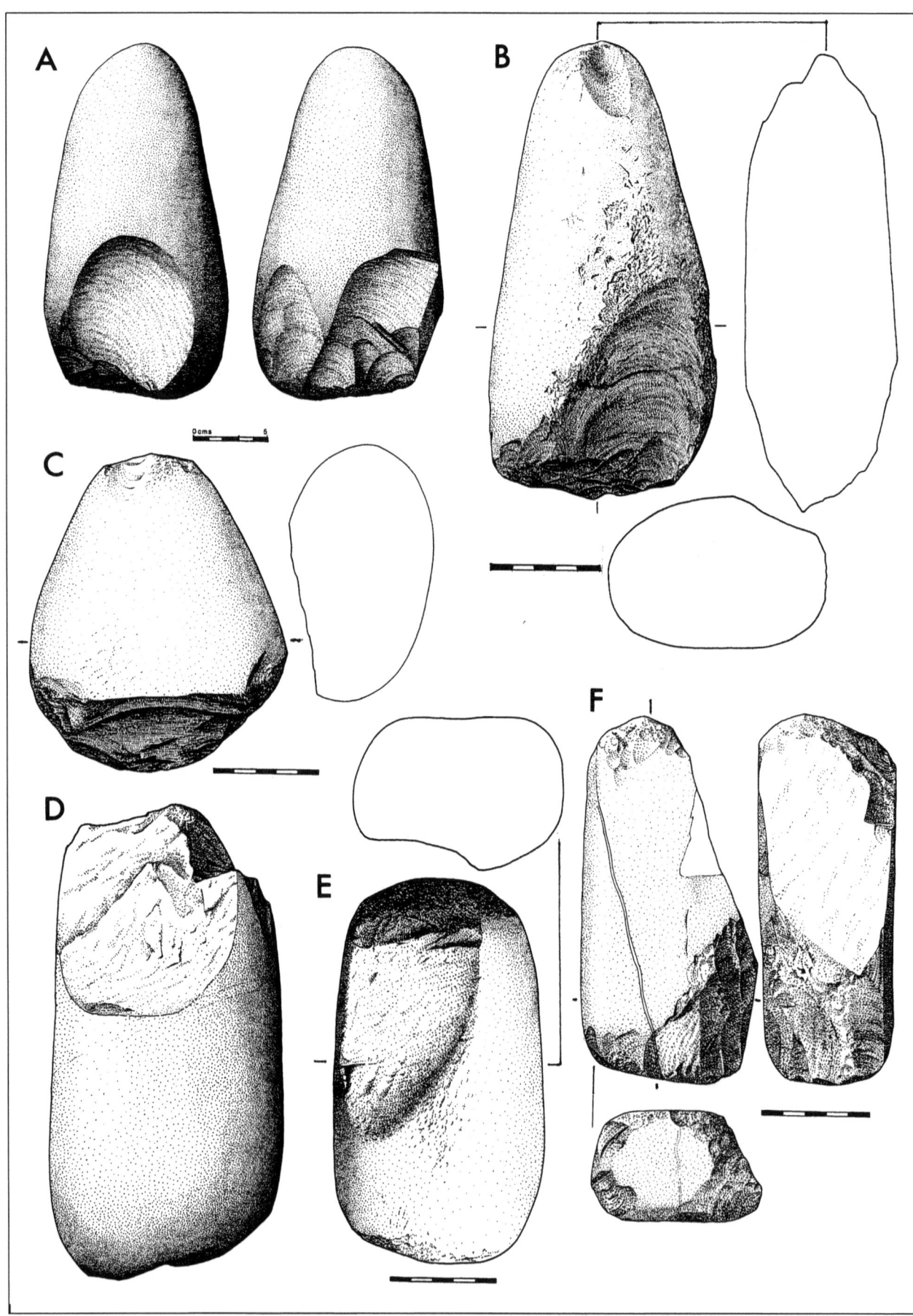

Figure 93: STONE TOOLS - End-hammers, picks etc (notched and semi-grooved)

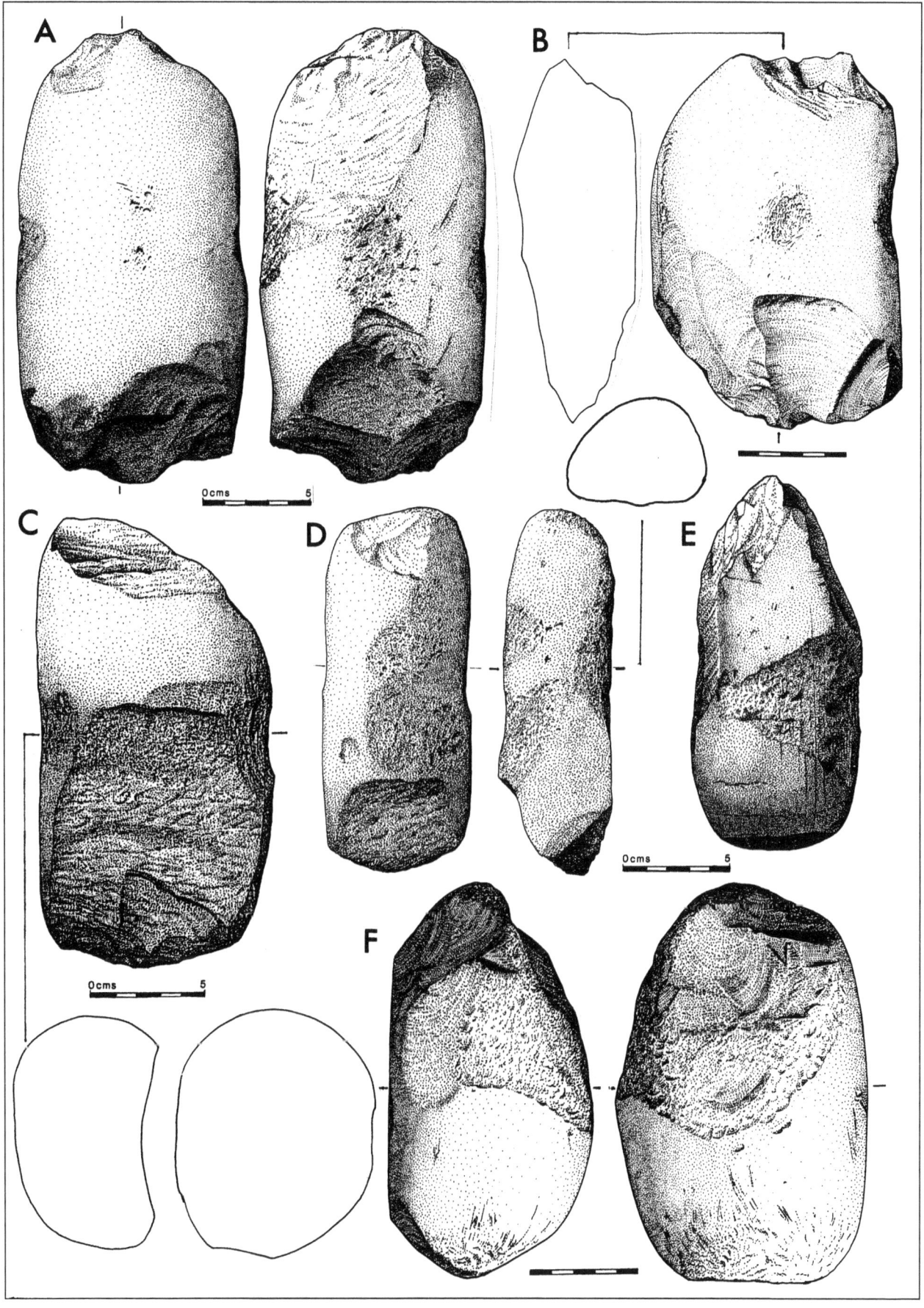

Figure 94: STONE TOOLS - Flake tools (chisels etc.), waste flakes and spalls

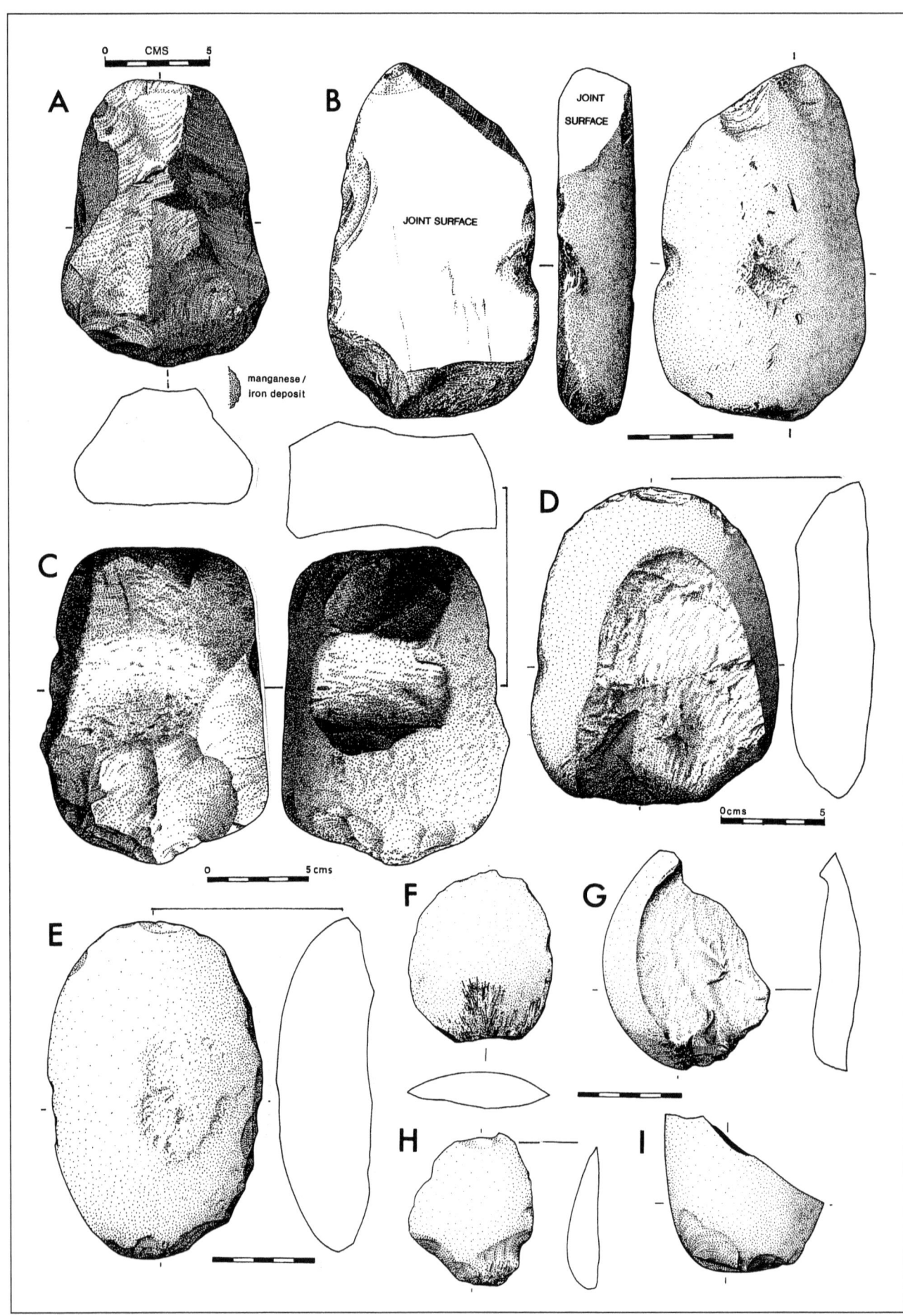

Figure 95: STONE TOOLS - Hand crushing/pecking implements

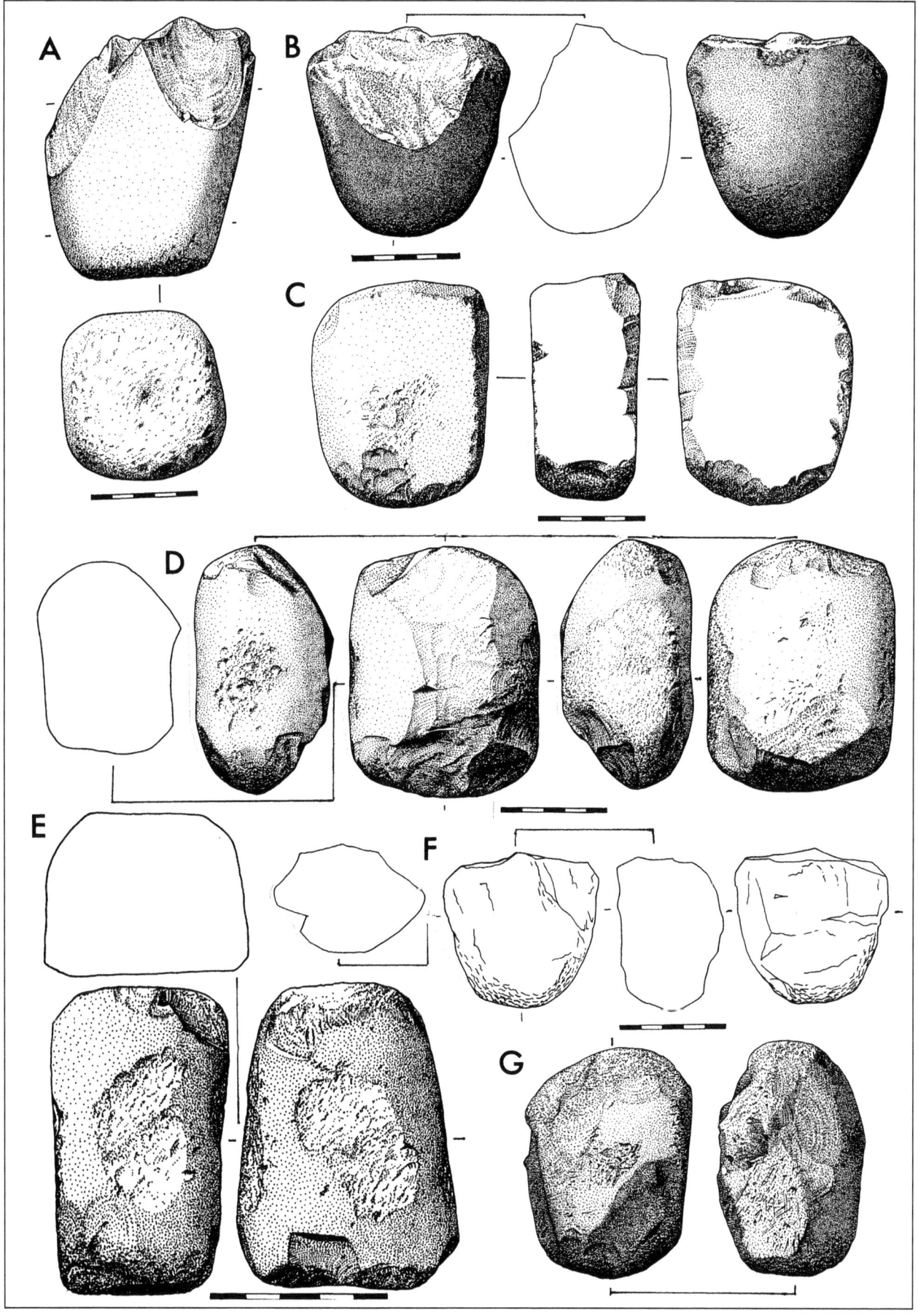

Figure 96: STONE TOOLS - Anvil stones, 'saddle-quern type' mortar, and miscellaneous (stone lid)

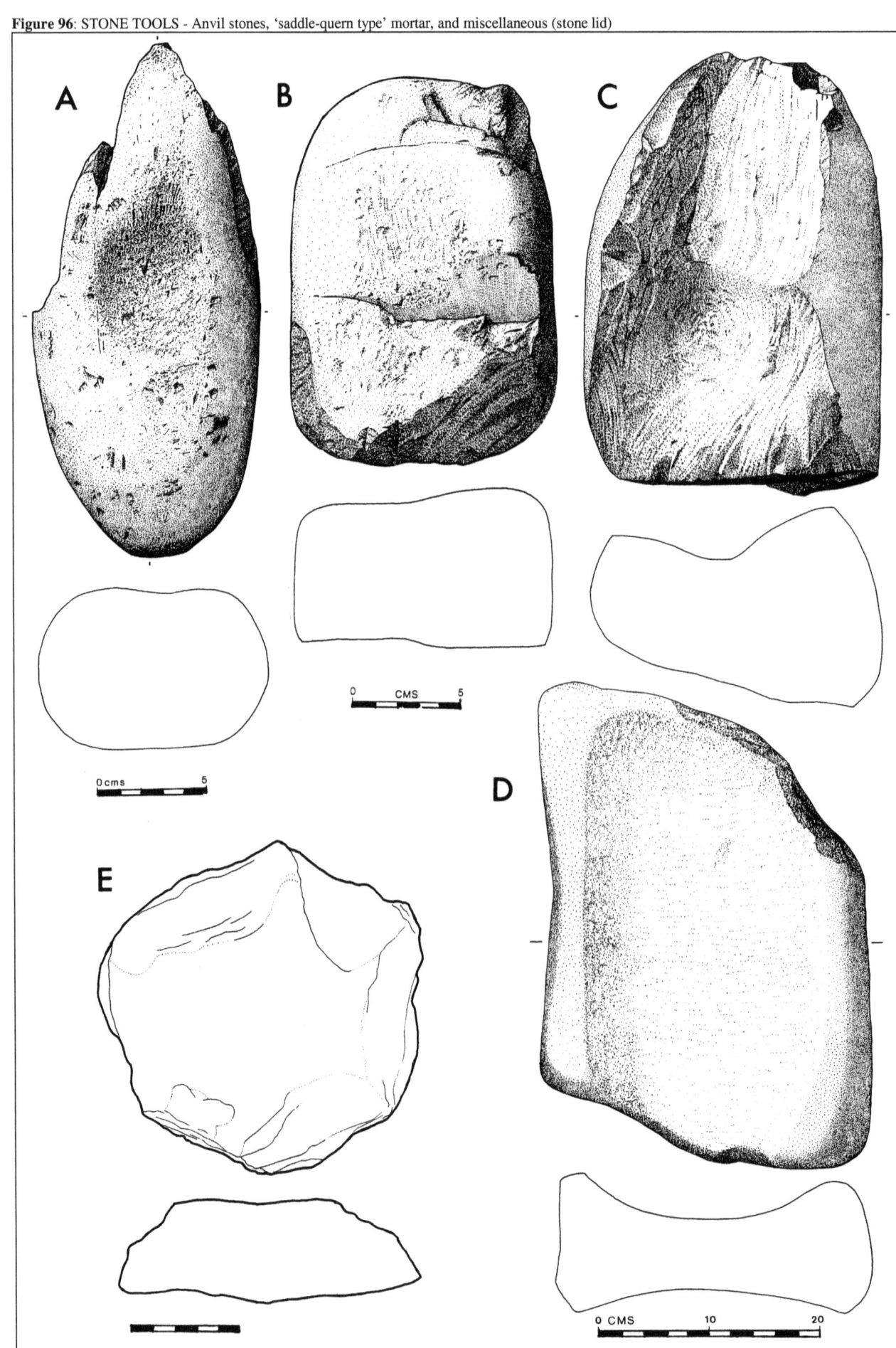

CHAPTER 15

ORE MINERALOGY (Rob Ixer, University of Aston)

The primary mineralisation at Copa Hill is simple and both the copper-rich, chalcopyrite-dominated and the spatially separated, lead-rich, galena-dominated ore assemblages carry identical mineralogies that only differ in the relative concentrations of their main sulphides. The present description, based upon additional sampling including *in situ* material taken from all seven of the cores recovered from drilling into the hanging wall of the vein in Fissure 1 using a hand-held rock corer (SEE Jenkins & Timberlake 1997,33-35) complements and amends earlier ones (Ixer 1997; Ixer and Budd, 1998).

The wallrocks hosting the mineralisation comprise fine-grained, slightly carbonaceous, meta-mudstones and meta-siltstones with abundant chlorite porphyroblasts. These metasediments are unmineralised but carry trace amounts of pyrite (FeS_2) some of which is framboidal and even smaller amounts of a white nickel-cobalt-iron sulpharsenide. The metasediments have been intensely silicified and brecciated so that many specimens comprise angular clasts floating in a medium- to coarse-grained quartz mosaic and later dolomite/ankerite matrix. These matrix minerals are the gangue to the base metal mineralisation.

The copper-rich mineralisation comprises chalcopyrite ($CuFeS_2$) accompanied by pyrite and minor amounts of galena (PbS) and sphalerite (ZnS). Trace amounts of marcasite (FeS_2) and a white cubic mineral that is optically identified as being one or more of the nickel-cobalt-iron sulpharsenide minerals belonging to the cobaltite (CoAsS)-gersdorffite (NiAsS) group or possibly ullmannite (NiSbS) are also present. Much of the sphalerite shows fine-grained chalcopyrite-disease and the nickel-cobalt-iron minerals show compositional and/or mineralogical zoning. Chalcopyrite is up to several centimetres in size but all other minerals are small, having a diameter less than one millimetre (1000 microns) across for pyrite, sphalerite and galena and less than 100 microns for the white sulpharsenide.

The primary copper-rich sulphide assemblage has altered/weathered to a more complex secondary mineral assemblage comprising iron and copper sulphides, oxides, hydroxides and carbonates. Although the alteration is seen to be widespread and quite intense in the spoil material, it is superficial in the *in situ* core samples. Chalcopyrite alters to a series of thin, copper sulphide rims including digenite (Cu_9S_5), covelline (CuS), yarrowite (Cu_9S_8) and spionkopite ($Cu_{39}S_{28}$) or to cuprite (Cu_2O) accompanied by native copper or to limonite (FeO.OH) and green copper secondary minerals including malachite ($Cu_2(CO_3)(OH_2)_2$). Alteration of the sulphides has produced intense de-dolomitization of the main dolomite/ankerite gangue so forming coarse-grained, vuggy calcite crystals. Limonite and black wad (pyrolusite (MnO_2)) infill these void spaces to give the dominant brown-black coloration of the specimens.

The lead-rich mineral assemblage comprises centimetre-wide galena masses intergrown with or enclosing small, less than one millimetre in diameter, chalcopyrite and sphalerite showing chalcopyrite disease, and trace amounts of pyrite and a white, unzoned, cubic mineral up to 50 microns in size and identified as ullmannite. Galena alters to dark coloured and then pale, euhedral cerussite ($PbCO_3$) or is enclosed within thin digenite, spionkopite, yarrowite or covelline rims.

Although a complete ore-triage (Ixer 2001) has not been performed on the mineralisation from Copa Hill it is possible to make some preliminary comments about the specimens as Bronze Age ores. The coarse-grain size of the chalcopyrite in the copper mineralisation suggests that a high grade, clean, copper separate could be achieved by hand cobbing. If smelted alone this would produce a copper metal with few other metallic impurities. These would include some iron from chalcopyrite, pyrite and limonite, trace amounts of lead and zinc from any admixed galena and sphalerite and perhaps a little manganese from the wad staining. Other elements namely nickel, cobalt, arsenic and antimony would have a small but negligible content, less than 0.01 wt %. Similarly a clean lead cut could have been collected from the galena-rich mineralisation by simple hand cobbing.

CHAPTER 16

ORES MINED IN PREHISTORY

MINERAL VEINS, ORE GRADE, AND THE EXTRACTION OF LEAD AND COPPER (Simon Timberlake)

At least five veins (A-E) associated with southerly downthrown fault fractures have been identified within the front of the opencast, all of which were tried in prehistory (Fig.97). The principle vein shoot (D) on the north side consists of a quartz-ankerite and rock breccia up to 2-3 metres wide, most of which has been removed within the working, leaving only sterile carbonate remaining upon the footwall beneath the prehistoric mine gallery. The base of this vein has been sampled by the postmedieval shaft, in the upcast from which could be seen ore containing numerous stringers of chalcopyrite, pyrite, blende and galena (within which the chalcopyrite makes up less than 5% of the vein mass). However, the grade of the ore removed in prehistory could have been much higher, leading to speculation about the percentage of copper present (Bick 1999; Mighall et al. 2000). Whether or not this rich ore shoot or pipe ever existed, it seems likely that some of the best ore values were still to be found here close to the surface, a factor which may well have led to the mine's discovery and perhaps also to the rapid rate of working during its first hundred years of existence. Even so, one can only guess at the grade and volume of copper ore present within this area. Worked to a depth of 15 metres from surface, a 15 m x 2 m ore shoot carrying up to 10 % chalcopyrite (average) may have yielded between 60 and 90 tons of copper mineral. Assuming chalcopyrite to be about 35% copper, this could have contained between 21 and 31 tons of metal. In reality, of course, the actual yield of metal was probably much lower than this, perhaps even when added to the remaining tonnage produced from all the other veins. The dilution of chalcopyrite with pyrite, as well as direct metal losses incurred throughout the process of ore concentration and inefficient smelting using primitive furnaces will all have been factors responsible in what may have been a much poorer recovery of metal.

For the most part the narrower veins infilling the faults B and C provide a much better insight of typical vein structure and composition at the mine. The worked-out areas of these consist of a fairly discontinuous quartz-ankerite breccia mineralisation which carries variable values of copper. This phase is only fully developed where the fault is open and cavity filled, for example at the change of angle, such as we find in the area of the Fissure 1 working. The latter seems to have been the deepest and almost certainly the next richest ore-producing area of the mine. This earlier phase lies against the footwall on the north side of the veins, whilst attached to the hanging wall is to be found a semi-continuous band of galena (between 5-10 cm thick) some of which was left *in situ* by the miners. Even where this fault becomes squeezed the latter can still be followed, and evidently the earliest miners used this knowledge to their advantage, following the fault in its barren areas, removing some of the lead and what showings of copper could still be gleaned through picking holes along the sides of the vein (SEE Fig. 53).

Better evidence for what appears to have been extraction for lead can be seen within the entrance cutting in the form of a narrow vein of galena (A) infilling a fracture which has been displaced by the fault at B, and therefore barren on its north side. Originally exposed by the cutting, the galena seems to have been removed to a depth of 10 cm from the vein in the floor, large lumps (between 50 - 500 gm) of this lying scattered around the surface of the rock either side of this, whilst the same vein on the west wall appears to have been much more thoroughly picked out, a thick layer (052) of finely crushed (but apparently unused!) galena lying upon the floor beneath this. Reflected light microscopy and SEM analysis of these galena fragments showed no evidence for any accompanying copper mineralisation, and the lead itself contained no appreciable silver content (*pers. com.* S.Dominy UCW Cardiff), suggesting that the removal of this could not simply have been for the extraction of associated metals/ores. Could the galena have been removed with the belief that this may have led to worthwhile copper ore beyond? Whatever the scenario with the extraction of lead ore (and there is no real reason to believe that this was ever carried out with the intention of smelting), the activity here seems to have been on a very small scale.

However, the incidence of discarded galena is considerably greater than that of visible chalcopyrite within the mine spoil, there being relatively few identifiable fragments of copper ore (including relict samples of ore now oxidised to goethite SEE Fig. 98) which larger than 5 - 10 mm in size, a fact which cannot simply be explained by the amount of oxidation and leaching that has taken place. Because of this particular attention was paid in 1999 to the recognition of unaltered or else oxidised copper ore within mine spoil, including the micro-excavation of small areas which may once have been crushing platforms located upon the summits or flanks of the tips outside the mine. For example, one lens of finely crushed mineral within an area of 0.5 square metres (Trench E6) revealed evidence for copper enrichment and also the presence of hundreds of crushed grains of chalcopyrite, most <3 mm in diameter and partly oxidized to goethite.

Figure 97: Conjectured plan of mineral veins and faults within the front-central area of the opencast. Drawing B.Craddock.

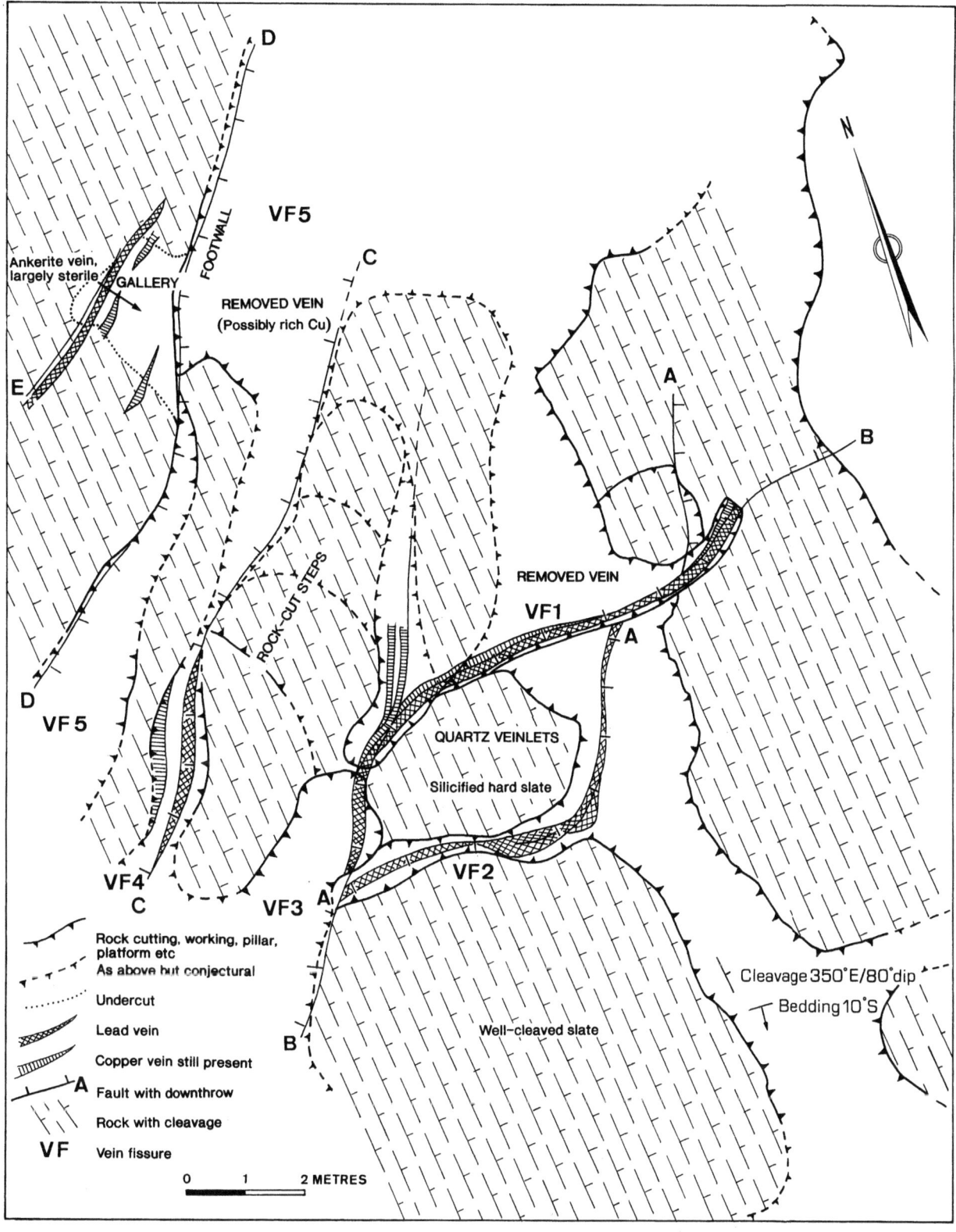

It has been suggested that most of these represent grains lost or rejected in crushing (Mighall et al. 2000). Similar results were obtained from a semi-quantitative mineralogical analysis of the <7mm fraction of weathered mine spoil sampled from the Central Tip. Geochemical analysis (carried out by ICP-MS) of all fractions of the same sample revealed actual metal levels of 4300 ppm Pb (lead) and 680 ppm Cu (copper), against 2400 ppm Pb and up to 5200 ppm Cu for a similar sized sample of mine spoil/processing sediment taken from within the mine itself (Jenkins & Timberlake 1997). The latter concentration may be due to metal precipitation upon the floor of the mine, or else to the much higher level of copper present wherever these layers are less affected by leaching. However, within the base of the mine entrance cutting, the alternate deposition of fine-grained waterlain and coarser grained crushed sediments, some of these richer in lead (052) than others (054/055), has been cited as evidence for mineral separation and ore washing, perhaps associated with the supply of water provided by the launder (Timberlake 1995). Unfortunately, such claims remain difficult to prove.

NOTES ON ORE PROCESSING (A. E. Annel, University of Wales, Cardiff)

Analysis of mineral grains, metal concentration and sedimentation textures found within different layers and fractions of recently excavated mine spoil was carried out by Dr.Alwyn Annels (Dept. Earth Science, Cardiff) in 1994. This provided additional information on the primitive methods of mineral processing. A further two sections (BC1 and BC2) were cut into the sides of the Central and NW spoil tips outside the mine.

Examination of the surfaces of crushed quartz and sulphide grains suggested that most crushing resulted from direct impact rather than rotational grinding (suggesting the use of stone-hammers on anvils), although some examples might indicate further comminution of the ore within mortars or querns. Nevertheless, most fractions of spoil were devoid of significant sulphide mineralisation, appearing to be either tailings or crushed waste (development rock), although some did show evidence for secondary enrichment by circulating groundwaters. Higher levels of lead compared to copper seemed to confirm that this metal was *not* being selectively extracted from the ore, the depletion of copper being due to its removal during processing. Where sulphides were present, the medium fraction (1 - 5 mm) contained by far the highest levels, suggesting perhaps that they reported mainly to this fraction during crushing, the greatest loss of chalcopyrite also being associated with this stage.

Figure 98: Polished ore (microscope) sections of chalcopyrite from the prehistoric workings: (**A**) grain of copper ore largely oxidised to goethite (brown) on exterior with relict of fresh chalcopyrite in centre (from prehistoric tips); (**B**) fresh chalcopyrite in quartz showing extensive brecciation.. From Entrance Area (A) inside mine; (**C**) remnant ore (now goethite grains) present within crushed <7mm fraction recovered from prehistoric tips. Photos ST 1996.

Figure 99: Small galena vein VF2 (shown shaded) excavated within rock floor at the north end of Entrance A. Lumps of discarded galena (Pb) lie discarded either side, whilst the stone chisel [CH94:h212] can be seen embedded in the crack. Photo ST 1994.

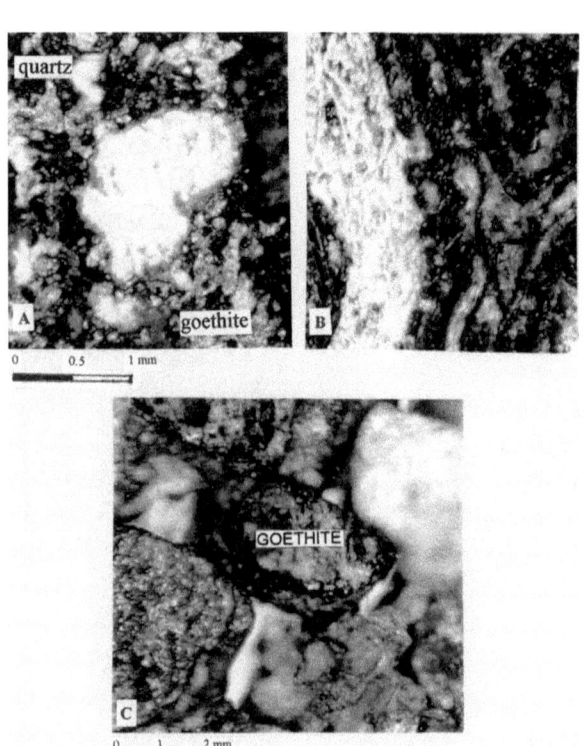

CHAPTER 17

A SUMMARY OF THE SEQUENCE OF EXPLOITATION AND ABANDONMENT HISTORY

Although this can be interpreted much more confidently in some areas and at some stages than others, it has nevertheless been possible to reconstruct a history of mining at this site from the end of the third millennium to the time of its abandonment sometime between 1500 - 1600 BC (Fig.100: A-I); a sequence followed by the natural infill, weathering, then finally the renewed prospection of the mine during the early modern period.

Uncertainty surrounds the earliest date for mining activity here, yet both topographical and archaeological evidence suggests that within the area of the present opencast the Comet Lode had been exposed by glaciation, and that the mineral vein(s) here outcropped as a small cliff face, some 2-3 metres high, with an accumulation of scree consisting of mineralized rock lying at its base. Thus the oxidation of the iron carbonate vein together with the green/blue staining of copper where springs issued at its base may well have been visible from afar, particularly if Copa Hill was approached from the south along the valley floor. Indeed, there is a suggestion of interest shown in the site some time before 3000 BC, perhaps for the purpose of collecting surface mineral for use as pigments (SEE A), although details of this still remain sketchy.

The earliest mining or prospecting activity may have commenced about 2500 BC. This took the form of surface collection of mineral followed by shallow pitting and the undercutting of the exposed vein in the cliff-face, with the creation of several small hillocks of spoil alongside (SEE B). Fire was used both to excavate the rock and break up the copper (chalcopyrite) ore, and there is also some evidence for stripping the turf from the site first.

Mining seems to have recommenced in earnest around 2000 BC following a period of abandonment (SEE C). The outcrop of the vein was attacked from the east side over a linear distance of about 20-30 metres, the blocky spoil being tipped to the side over the earlier hillocks, thus forming a continuous ribbon of spoil (the Lateral Tip), as this vein was removed horizontally and progressively downslope. Deepening of the working resulted in the formation of a trench as the main ore-shoots on the eastern side began to be mined up-slope from the south end. With no proper level entrance and with deeper pit-like workings in the centre of the trench the miners must have begun to encounter water problems. At the same time, the retreat of the low cliff westwards (as this eastern ore-shoots were excavated away) resulted in the discovery of the main vein.

Between 2000 and 1950 BC mining continued upon the east and west sides (SEE D) although the main working was by now from the front (on the east side), reaching a depth of over 2 metres within the sulphide zone, and encountering lead and zinc as well as primary and secondary copper ores. Most of the veins at the south end had already been discovered, yet work still concentrated on the east side, the veins now being worked by a series of interconnecting trenches. Initially the main entrance would have been somewhere in the middle at the front end of the working, although the entrance cutting at the SE end was probably cut shortly afterwards. At the beginning of this period the system of launders may already have been in use; diverting water away from the springs which were even then threatening to drown the working. These drains were probably used at different locations within the mine, and thus were moved around as the workings got deeper.

Considerable development seems to have taken during the period 1950 - 1900 BC. Deeper mining continued on the south and eastern sides, coinciding with the maximum excavation of Fissure 1, vein B, and the completion and first use of the entrance and drainage cutting A within the south-east corner. Exhaustion of the reserves of copper on this side of the trench is also suggested by the excavation into and subsequent abandonment of a lead vein (Fissure 1). Considerable tipping downslope of the opencast continued from the eastern side (SE Tip) of the main vein workings, then moved westwards (Central Tip [SEE Fig.101 sequence 4a-c])) as the new entrance cutting became a route for the removal of spoil. Deeper mining was also commencing on the main vein (C) on the north-west side, with an independent entrance to this working from the front, as the vein began to be sampled downslope of the main working by means of shallow opencuts.

The mine reached its maximum depth (10 m +) sometime during the next 50 years, perhaps between 1900 - 1850 BC (SEE E). Exploitation of the main vein continued along the northern side whilst a number of much smaller lead and copper veins were also tried - and the entrance to D9 (along vein B) was widened and deepened. Access into the deeper workings would then have been via a series of stepped rock benches with pillars in between, water being raised by rope and skin bags from a sump which had been allowed to accumulate at bottom of the mine up to the level of the entrance cutting. From this point it was carried away using the main (alder) launder(s). By the end of this period serious water problems must have rendered the bottoms inaccessible, a situation aggravated by the seasonal or even intermittent nature of the working. Because of this, an aqueduct system for diverting water may still have been in use. Spoil continued to be tipped downslope outside the mine (Central and SE Tips), whilst these areas as well as the floor of the entrance cutting and tops of the rock benches were also being used for the crushing, grinding, and hand sorting of ore.

The deepest parts of the opencast, particularly on the east side, appear to have been abandoned by about 1840 BC, and thereafter remained semi-permanently flooded up to a level 1-2 metres below the floor of the entrance, with redundant woodwork including several launders and mine timbers being jettisoned into some of the disused workings (Fissure 1) [SEE F]. It is possible however that the water level was lowered for short periods of time on re-commencing work. This may have coincided with a temporary abandonment of the whole mine sometime during the period 1850 - 1800 BC, although it seems much more likely that an intermittent, but nevertheless significant amount of mining was still taking place, such as within the area of the perched mine gallery and north-west vein shoots, and perhaps also up-slope towards the shallower northern end of the trench. Some of the spoil was now being tipped inside (into the abandoned southern and eastern vein workings), although most was still being carried out of the mine, perhaps because the miners were aware that ore reserves continued at depth and remained hopeful that the bottoms would eventually be drained.

Around 1800 BC it seems that there may have been a renewed attempt to work the principal vein in front of the opencast through trenching downslope on the Comet Lode, tipping spoil from this to the north (NW Tip) [SEE G]. Indirectly this would have helped to lower the water level within the mine itself via opening up a slightly deeper entrance once more on the north-west side. Meanwhile periodic mining and prospection work continued at the far end of the opencast where the lode values pinched out. This can be traced in the fresh deposition of mine spoil beyond the northern end of the old Lateral Tip. Before or during this period (1800 - 1700 BC) the SE entrance cutting appears to have become blocked, yet the launder appears to have been dug out on several occasions, attesting to continuing activity, possibly for drainage, but perhaps also the crushing, washing and separating of a stockpile of already mined ore (SEE H). Some of the implements of antler and small withy appear to date from this time. The latter work seems to have continued intermittently for many years, as evidenced by the layers of charcoal tipped over the launder, and finally the blockage of this entrance, either through the natural slumping of tip material, or else as a result of its re-deposition following an attempted re-working of the lode or spoil tips for copper and/or lead during the period circa. 1650-1600 BC. During this latter time no activity appears to have been taking place within the interior of the opencast, and rapid silting-up of the drowned working ensued.

The mine effectively now lay abandoned, and natural infill of the opencast followed. A short-lived final attempt to clear the blocked entrance cutting, perhaps to re-activate the drain and lower the water level, occurred circa. 1550 BC. However, this trench failed to reach or to remove the remaining launder, although the level of dammed-up water within the working may have been lowered by up to 1-2 metres. There is also some suggestion of further movement of spoil, or perhaps even renewed tipping outside the mine, although the source or purpose of this remains a mystery. This may represent small scale re-working of the tips for lead (I).

By 1400 BC the centre of the opencast was almost completely infilled, the boggy surface of this lying at a depth of only 1-2 metres below the rock lip. Some two to three hundred years later the climate had further deteriorated, yet the top of the short mine gallery against the north wall of the working was probably still open, with leaves fallen from the trees now overhanging the sides of the opencut gathered at its base. Despite there being some evidence for agriculture and woodland management on the valley sides, there were now no traces of mining anywhere within the vicinity.

LATER HISTORY

Although some Roman mining activity is indicated on Copa Hill (SEE Mighall et al 2002 b), the opencast itself appears to have remained untouched. Later, during the 11th-12th century AD the interior of the opencast was re-examined by miners who sank a shaft, which was probably abandoned shortly afterwards. This was followed by the construction of a leat from the upper reaches of the Nant yr onnen to carry water to flood the infilled opencast depression for use as a hushing pond. Hushing could have been used to excavate the earlier spoil tips on Copa Hill in order to recover lead, but equally it may been to uncover the Comet Lode downslope. The latter work was probably carried out by or on behalf of the Cistercian monks of Strata Florida.

During the 18th century (c.1730) further prospection work seems to have been carried out in the opencast, including the sinking of a shaft followed by its rapid abandonment. Subsequently (1813 or before) a larger shaft was sunk through the base of the opencast to sample the unworked lode, meeting a rise driven from the Copper Level some 27 metres below. This shaft drained the base of the prehistoric working, but otherwise caused little disturbance to these 4000 year old mining deposits.

Figure 100 A-I: The sequence of prospection and mining through to abandonment of the prehistoric mine (c.3000 - 1500 BC). Drawings B.Craddock.

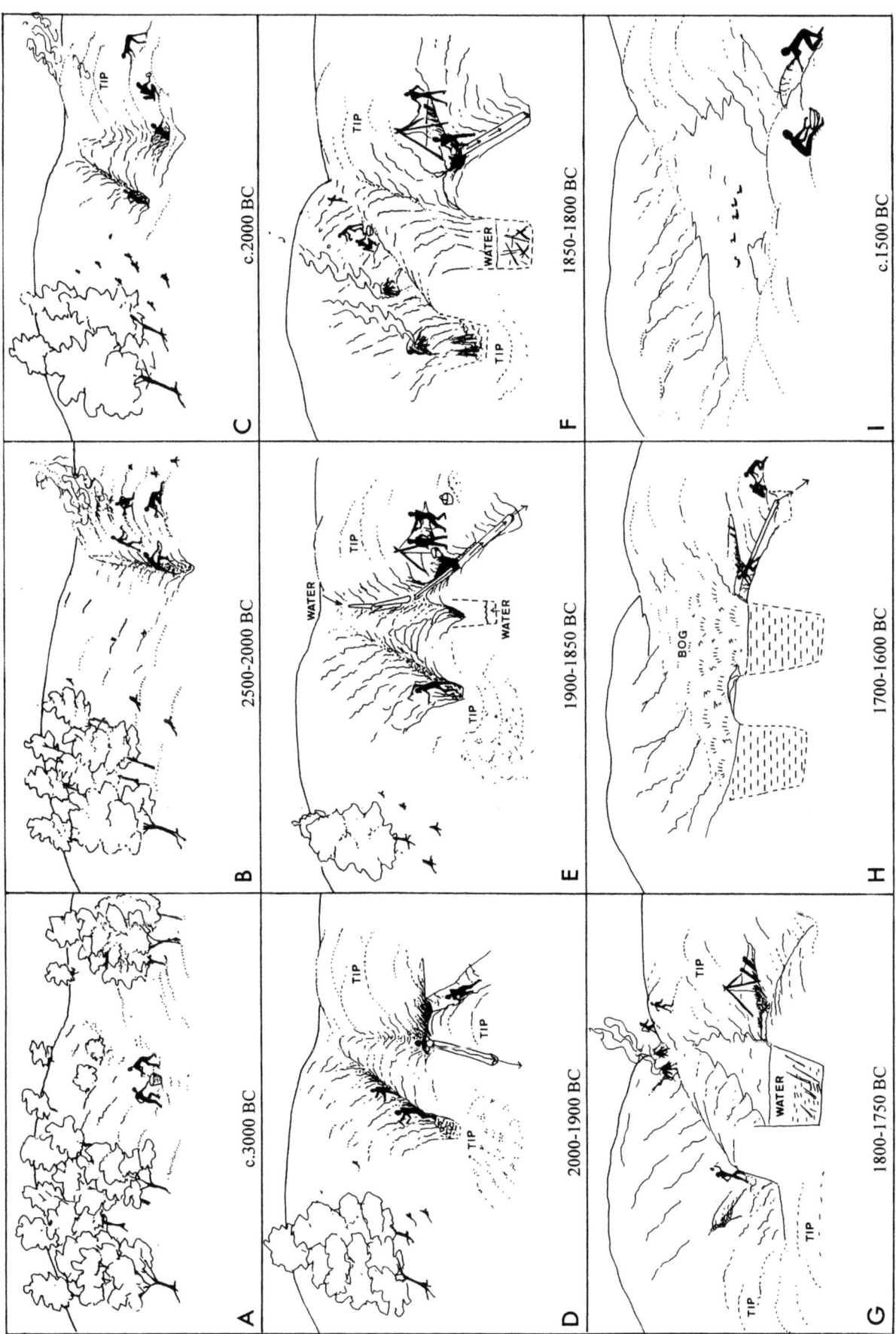

Figure 101: Conjectured sequence of tipping outside of the prehistoric mine - commencing with prospection hillocks (1a), Lateral Tip (1b) spoil downslope from initial trench (2), earlier phase of tipping eastwards downslope (3a-b), main phase of tipping centrally (4a-c), first infill of Entrance A (5), NW Tip (6a-b infilling former surface opencuts), late working at north end of opencut (7), cut of late channel and re-working of tips (8)

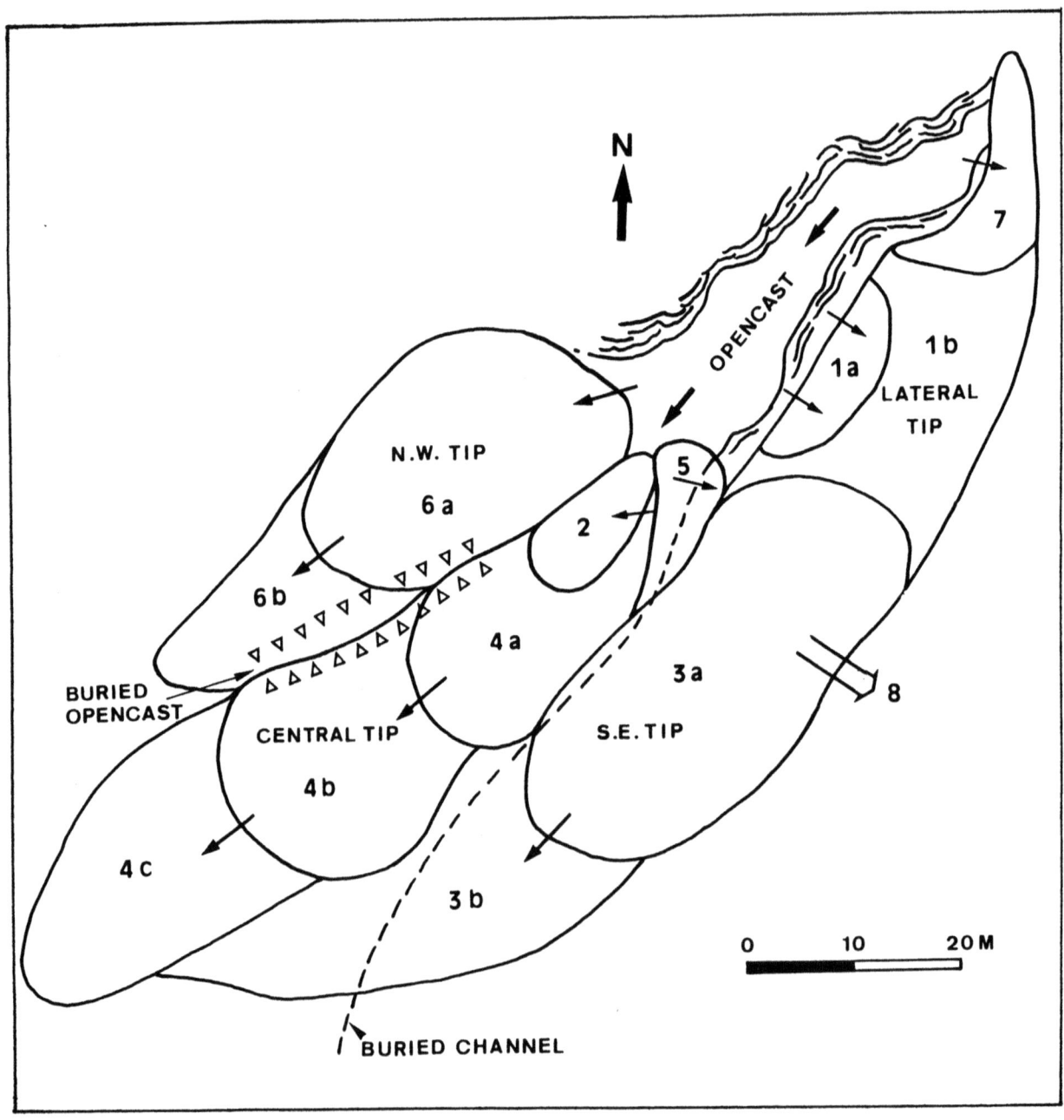

CHAPTER 18

DISCUSSION

The scale of mining

Comparison of the various different estimates for production from Bronze Age mines within the British Isles makes interesting reading, yet it is a frustrating exercise, particularly when one realizes that almost all of this comes from mines which were most active during the earlier part of the period, and which had apparently ceased production by the Middle and Late Bronze Ages, at a time when the use and availability of metal was many times greater (Pearce 1984). For example, the actual sources of Welsh copper used in the Acton Park industry which it is claimed supplied much of Southern Britain with metal during the Middle Bronze Age (Northover 1982; O'Brien 1999 - Der Anschnitt) have still not been identified. Yet earlier claims for production from Mt.Gabriel exceeded 146 tons of copper metal, with a total of 372 tons for SW Ireland as a whole during the Early Bronze Age (Jackson 1980). These figures have now been substantially revised downwards by O'Brien (1994, 196) to a minimum of 2.94 and 1.47 tons of copper for all 31 of the Mt.Gabriel mines, assuming a total extraction of some 4000 tons of rock and a 10% smelting efficiency and recovery rate from separated ore (a maximum of 26 tons would therefore be possible at 90% smelting efficiency). This contrasts with a still more conservative estimate of only 0.6 tons of copper metal smelted during the entire Irish Early Bronze Age (Flanagan 1980). We find rather similar problems concerning estimates for production at the Great Orme Mines. For the latter site these figures (likewise critically dependent upon actual ore grade and metal yields) range from 175 to 238 tons of copper metal recovered from an estimated 41000 tons of ore, most of which was probably extracted before 1500 BC (Lewis 1996). At first sight, the latter amount seems perfectly credible when one calculates the size of the opencast and underground workings and views this in terms of modern demands for metal and current recovery rates, yet it begins to make less sense as soon as one starts to compare this with the total volume of metalwork recovered from the entire British Bronze Age, then taking into account loss, calculates the probable amount *actually* produced during the earlier part of this period. Thus these substantial claims for the Great Orme contrast with an estimate of only 6- 10 tons of copper produced from all of the mines in mid-Wales during the Early Bronze Age (Timberlake 2002), of which Copa Hill may have contributed no more than 2 to 4 tons (Timberlake 1990b). This latter figure was calculated upon the basis of the tonnage of spoil present within the tips outside of the mine (4800 metric tonnes), the assumption being that the vein here carried only small pockets of chalcopyrite (at 20-30% copper), with good ore being less than 1% of the total rock extracted, this being coupled with a minimum recovery rate of 30-40% after hand-picking, crushing and smelting. More recent calculations based on a much better understanding of the internal relief of the opencast (assuming a depth of about 10 metres at the front) has come up with a somewhat smaller figure of between 3300 to 4000 tons of rock removed, a result which leads us to suspect that this working could be a lot deeper than previously thought (perhaps 30 metres or more in places?). The presence of narrower but possibly richer ore shoots may also imply a higher ore yield, perhaps exceeding 100 tons of chalcopyrite. Of course, the true figure may well have been less, and even then, perhaps only a fraction of this sum was ever successfully smelted. Looking at it this way, ore tonnage could have been higher but the actual metal yield could still have remained the same (<4 tons).

In the absence of any clear archaeological evidence for smelting within the vicinity of the mines (either at Cwmystwyth or elsewhere) the production of metal and thus the technology of indigenous metal production continues to remain a matter of considerable speculation.

Where is the smelting evidence?

Despite the evidence for atmospheric pollution during the mining period, some of which could be attributed to local smelting, not a single fragment of scoria, slag, metal prill or furnace lining/crucible has ever been recovered from the area of these workings on Copa Hill. However, recent evidence of Bronze Age smelting from Pen Trwyn on the Great Orme (Chapman 1997; Jones 1999) seems to suggest that this activity could have been undertaken some distance away from the mine. Moreover, such furnaces may have been small, simple and impermanent, all traces of which were extremely localised. For example, most of the small volume of slag produced may have been finely crushed in order to remove entrapped prills of copper. Furthermore, the rapid disintegration of the poorly fired clay lining of these sorts of furnaces has also been suggested following the examination some 18 months afterwards of the site of a small experimental bowl furnace used for the smelting of tin ore at Flag Fen (Timberlake 1994 c). Here, and at sites subsequently used for the experimental smelting of copper ores (where short-lived temperatures in excess of 1100° C had been reached), there remained little evidence for any metallurgical activity at all. Such a situation may have been the norm rather than the exception with the poorly reducing smelting furnaces typical of Early Bronze Age Britain. Craddock (1995) suggests that this primitive smelting technology may also have inadvertently used what was essentially a non-slagging process, another possible reason for the absence of slag from mines or settlements of this period. Nevertheless, it is generally accepted now that sulphides, including both fahlerz and chalcopyritic ores were being exploited both in Ireland and in Britain from 2200 BC onwards (O'Brien 1999; Northover 1999). Indeed it has been shown experimentally that it is possible to smelt chalcopyrite under poorly reducing conditions; initially by roasting this in heaps at 700° C, followed by heating the crushed part-roasted ore within crucibles under charcoal at about 1250° C (Rostoker et al. 1989). The latter method will produce prills of copper inside of a viscous spongy mass, a true slag forming only inside of a proper shaft furnace on using a more impure (iron or silica rich) ore, and/or with the addition of quartz. Thus the recent discovery of substantial 'roasting pits' at the Beaker period mining camp on Ross Island (O'Brien 1998) as well as the funnel-shaped smelting hearth at Pen Trwyn, Great Orme within which chalcopyrite was apparently smelted (*pers com.* Peter Northover) would seem to imply a rather similar sort of primitive smelting technology – one which might also have been practised in mid-Wales. Budd et al. (1992) propose an even simpler scenario of poorly reducing 'bonfire furnaces' (at temperatures of 700° C +). In these it would have been possible to smelt mixtures of green oxidized minerals including malachite and copper arsenates, but not however the sulphides, which it now seems that most of these mines were busy producing. Furthermore, the complete absence of smelting traces at the vast majority of these mines suggests that this activity must primarily have been linked with the associated work camps or settlement sites, all which still remain to be discovered. Within the upland zone, particularly with hill-top sites such as Copa Hill, it is easy to understand why such settlements might need be located away from the mine, situated instead close to sources of fresh water, fuel and the better grazing land available on the valley floor or lower valley sides, sites to which the few bags of hand-picked ore produced at the mine could be brought at the end of each working day (Timberlake 1990 a). Indeed, this seems to have been the situation within the more important Alpine mining area of the Mitterberg in the Austrian Tyrol (Pittioni 1951; Shennan 1999). A rather similar model with separate mining, processing, and smelting areas also seems to have been true of lowland sites during the Chalcolithic - Early Bronze Age, as we find at Cabrieres, SW France (Ambert et al. 1984) and Aibunar in Bulgaria (Cernych 1978). Perhaps the ore concentrate itself could have been a product traded and exchanged by miners? In this way the smelting of the ore may have ended up being carried out some distance from the mineral source.

The early exploitation of lead at Cwmystwyth

Whilst there has been little in the way of general discussion over the earliest use and exploitation of lead within Bronze Age Britain, it has been accepted almost without comment that lead was being intentionally added to bronze both as a dilutant and to improve its casting properties from at least the beginning of the Middle Bronze Age – in this case the Acton Park II metalwork production period (Tylecote 1986, 30), the suggested origin for much of this metal being North Wales (Northover (in Savory 1980, 233)). However, no specific sources for the ore itself were suggested, and as yet no obvious links have been made with the very meagre finds of stone mining tools discovered at what appear to be exclusively lead mining sites, such as at Craig y forwen nr. Llangollen (Frost & Hankinson 1995) or Wagstaff Lead Vein on Halkyn Mountain (Walters 1994). It is interesting therefore that Copa Hill has provided us with some of the best evidence yet for the extraction of lead ore some 300 to 500 years before we find evidence for its (still uncommon) use in metalwork. Moreover, the issue of whether or not its removal was being carried out wholly or partly to obtain the copper ore associated with it, or whether this was also to supply an existing small-scale demand for lead or silver (as yet unrecognised during the Early Bronze Age), or else simply to foster an interest in 'metallurgical experimentation' (galena being one of the most easily smelted sulphide minerals (Tylecote 1986, 54)) continues to be the subject of much eager debate (Bick 1999; Mighall et al. 2000). Although the balance of evidence still seems to suggest that copper was indeed the primary goal of the Cwmystwyth miners, Bick is quite right to point out that almost all of the veins prospected or worked by Bronze Age peoples in mid-Wales were primarily lead-bearing, at a time when many of the much richer copper deposits which lay to the east in Montgomeryshire and northwards in Merioneth and Caernarvonshire were apparently being ignored. One possible explanation is that those deposits containing a mixture of both copper and lead minerals were being preferentially selected, although the real reason for this, if indeed it had a metallurgical basis, is far from clear. Indeed it is hard to completely ignore the possibility of some link between this early association of lead and copper ores and the several repeated phases of intentional or even unintentional additions of lead to bronze seen during the Middle-Late Bronze Age. And despite the apparent purity of Early Bronze Age copper and the fairly uniform composition of tin bronze (Rohl & Needham 1998; Northover 1980) should we not be seriously looking for more evidence of early lead artefacts, for signs of metallurgical experimentation, or even traces of lead and silver as contaminants within bronze?

Smelted lead first appears during the 7th millennium BC in Anatolia, but this was not widely exploited till the late 4th/ early 3rd millennium when its use was expanded in the Mediterranean and Near East because of its role in silver production (Gale & Gale 1981). However, still earlier evidence for experimentation in smelting lead minerals can be found within the Balkans at Vinca culture sites of the 5th. millennium BC (Glumar & Todd 1987). By 2700 BC lead ores were being mined and lead artefacts were being produced by the Nuralgic peoples of Sardinia. Thereafter the fabrication of lead beads can be traced through the late Neolithic and then into the Chalcolithic within Languedoc, SE France (Arnal et al. 1979), whilst the earliest evidence for the use of galena in Britain (un-smelted as strung beads) comes from a Neolithic chambered tomb at Quanterness on Orkney (Henshall 1979). Artefacts of lead metal however are much rarer. The few known examples include an Early Bronze Age pin with a tin-lead alloy head from St.Columb, Cornwall (Shell 1979), an urn encrusted with lead foil from Sheury, Co.Tipperary (Rafferty 1961), plus the recent find of a lead bead necklace from a cist burial at West Water Reservoir in Peebleshire, the trace element composition of which shows a best match with a source of low-silver galena located at Silber Law on the Pentland Fault (Hunter & Davis 1994). It was suggested that the small scale exploitation of this may have been linked to local prospection for copper sulphide ores, hence experimentation with a local resource. At Alderley Edge, Roeder & Graves (1905, 8) observed that the Bronze Age miners had removed blocks of galena from their pit workings, but that this had not apparently been made use of, other than to form fireplaces, possibly for the roasting of copper ores. Today one might wish to re-interpret such widespread evidence for calcined galena as indicating some form of metallurgical experimentation, or attempts at smelting.

Notwithstanding the increasing use of lead in bronze towards the end of the 2nd millennium (Acton Park, Ewart Park and Wilburton industries), actual artefacts of lead remain rare throughout the British Bronze Age. Thus the discovery of a socketed axe of lead from Mam Tor, Derbyshire was also interpreted as being the result of 'experimentation', constructed perhaps for use as a lead pattern to make clay moulds as has been suggested of examples from Lincolnshire (Tylecote 1962), yet this former may represent the very earliest example of exploitation of the lead resources of the Peak (Guilbert 1996). Similar socketed axes of almost pure lead from Isle d'Est in Brittanny have also been interpreted as local products (Penhallurick 1986, 93). From Britain, Late Bronze Age bronzes containing up to 15-20% lead are all thought to have an indigenous source (Northover 1980 a). For instance, based on lead isotope evidence, Rohl and Needham (1998) considered Alderley Edge to be one of the possible origins for some of the lead and copper used in Ewart Park IMP LI-21 metal (1020-800 BC), whilst lead from the Mendips may also have been used within the Wilburton metalwork. Todd (1996, 50) suggests pre-Roman working of Mendip lead based on artefact finds (late 1st millennium BC) from Charterhouse. Meanwhile a Central Wales source has been suggested for one of the latter (Late Bronze Age) metal types (IMP-LI 17), though the impurity pattern within this fails to match the galenas either from Cwmystwyth or from any other of the known mines. Indeed, at the present time there are no clear matches between *any* Bronze Age artefacts and the galena extracted from Copa Hill.

Copper and Early Bronze Age metalwork within Wales and beyond

Very similar problems attend the matching of copper and bronze artefacts with known sources of copper metal, and thus by inference ores. The exploitation of as yet unrecognised mines within the copper-rich regions of SW England (Devon and Cornwall), Snowdonia, Cumbria, and SW Scotland has regularly been invoked to explain the wide range of existing metal types (Northover 1999), yet it seems strange that many of those sources already identified as mines (such as the Great Orme) were themselves capable of producing sufficient quantities of metal to meet almost all the demands of indigenous metalworking industries operating within the period 2000 - 1600 BC. Prior to this, the provenance of Northover's Type A copper metal, with an impurity pattern characterised by arsenic, antimony and silver characteristic of Irish Copper Age metalwork (also recognisable within some of the earliest flat axes from Wales such as the Moel Arthur hoard (Northover 1999)) can be linked with some confidence to the *fahlerz* tetrahedrite-tennantite and chalcopyrite ores of the type exploited at the Beaker period copper mine of Ross Island, Killarney (O'Brien 1998 & 1999). This has been confirmed by lead isotope analysis, the results of which show a good correlation between the metal type IMP-LI 1 present within much of the latter metalwork and the Ross Island orebody, there being some, but few other sources being exploited prior to 2100 BC (Rohl & Needham 1998). The succeeding Brithdir and Mile Cross assemblages (IMP-LI 4 & 5) may also include Irish metal, but in addition these suggest a plurality of other possible British source deposits including SW England, Scotland and Wales (the Great Orme). Imports of Irish metal (Type A) appear to cease altogether by about 1800 BC (Northover 1999), by which time the first indigenous bronze using industries in Britain were already making a significant contribution to the available metal pool. Wales was one of the main production centres, and it is here that we find a diversity of regional metal compositions. This includes Northover's 'C' metal (without any major impurities) centring upon North Wales and the Marches, with the Great Orme as a possible source, plus types B1, B3 and B4, all with variable amounts of arsenic (between 0.2 and 1%) and nickel (0.05 - 0.1%), centring on North Wales, mid-Wales (B4) and South -East Wales respectively (Northover 1980 b). Within the British context the earliest part of this bronze using phase is contemporary with rich Bush Barrow grave series (Wessex I), which in terms of metalwork corresponds to the Willerby assemblage dating to 1900 - 1700 BC and the metal type IMP-LI 6 (Rohl & Needham 1998). The latter is characterised by low impurity metal (arsenic:nickel <0.14%; antimony:silver <0.08%) of similar composition to Northover's B3 and B4 , a type which also shows a reasonable match in terms of its lead isotope ratios with the cluster of copper ores from Central Wales on the England and Wales Lead Isotope Outline (EWLIO). Despite considerable overlap between the Central Wales and other regional ore clusters, Nantyreira and Cwmystwyth both show a good correlation with IMP-LI 6, particularly during the earlier part of this period (Rohl & Needham 1998, 91). The spread of isotopic compositions may reflect two or more sources for the lead ore contaminants, possibly as a result of sampling different phases of mineralization present at the same locality (such has been identified within the Comet Lode at Cwmystwyth (Mason 1994)), but much more likely to be due to the mixing of metal from different geographic areas, such as the orefields of North-West Wales or Shropshire. The low impurity levels associated with IMP-LI 6 metal are perhaps significant. Fairly constant low silver, a stable element in remelting, implies mixing with other low impurity metal, whilst occasionally high lead present within some of the

Willerby metalwork (such as in the groups from Lydd and West Overton) could reflect accidental inclusion where lead ores are closely associated with copper. The latter is a situation analogous to Central Wales, but it should be said, to many other orefields as well.

So far the 'best fit' in terms of metal composition, isotopic signature and chronological date between Cwmystwyth ores and Bronze Age metalwork appears to be with this earlier part of the Willerby period, yet a match has also been suggested on isotopic grounds with the final Penard/Wilburton phase (1100-1000 BC) of the Late Bronze Age, although it seems that most of the ores used in the latter case were fahlerz (Rohl & Needham 1998,181). On this basis and the dating evidence alone it is doubtful whether Cwmystwyth could ever have been the origin for the latter metal. In future more accurate provenancing will come from the refinement of the techniques for trace element analysis of copper and bronze artefacts and ores, including the analysis of microscopic sulphide inclusions remaining within the smelted metal, and perhaps also the calculation and comparison of the ratios for what may be the most reliable differentiator elements e.g. Ag:Cd:Sb, Ag:Co:Ni, and Ag:Ge:Ni (Jenkins & Timberlake 1997, 42), whilst experimental smelting work will also help us to better understand the thermodynamics and products of early smelting furnaces, hence the eventual partition of impurities between ores and metal.

An examination of Early Bronze Age metalwork distribution within Wales shows no evidence for regional lusters of finds centred upon the mining areas of Central Wales, nor of a fall-off in distribution away from them, and thus it has been said that there is no evidence here for 'production zones', as is the case with pig lead ingots and Roman mines (Briggs 1988 & 1994). Equally there appears to no evidence for an increased density of metalwork around the presumably much larger Bronze Age mining centres of Parys Mountain and the Great Orme in North Wales. O'Brien (1990) noted a similar paucity of metalwork close to Mt.Gabriel phenomenon, which he suggested could be due the smelting of ore: away from the mines, such as around the coastal lowlands of West Cork. Alternatively thw ore may have been traded still further afield for treatment, thereby establishing a link between metalwork and secondary distribution centres rather than with the original mined source, as has been suggested for British stone axe production in the Neolithic (Cummins 1979). Judging by the widespread distribution of Irish type metal throughout Britain at the beginning of the Early Bronze Age, it would also seem likely that the trade networks for copper of Welsh origin were equally far-reaching, the alloying and production of artefacts (eg. axes and halberds) taking place some distance from the mining centres. Following the further mixing of copper, its alloying with tin, then re-melting for casting in axe moulds it is quite possible that the bronze smiths themselves were unaware of the exact source of any metal (Timberlake 1992 & 2001). Such a situation may well have applied to Willerby or other Early Bronze Age metalwork.

The latter consisted mostly of axes, with a trend in axe decoration which reached a peak of complexity around the Willerby/Arreton transition c. 1700 BC (Needham et al. 1985). Typical of the earlier forms were the development of more crescentic (convex) cutting blades c. 70-120 mm wide, featuring low flanges and stop bevels 1-1.5 cm deep, the latter which would have produced stop-marks in tooled wood of about 7-8 cm +. This is interesting in respect of the size and shape tool marks recorded within the wooden launder recovered from the mine, clearly indicating manufacture by a flat axe with a blade of this general type. Although the apparently earlier date of construction this launder is a problem, its use evidently continued into and throughout the above period. However, a more reasonable explanation is that the sort of axe used was of rather similar design, but one which pre-dated (c. 2000 BC) the most significant production of metal from Cwmystwyth or the other Welsh mines. In this respect a comparison with a more crescentic bevelled blade of Brithdir design seems more appropriate. Similar problems matching the metalwork/ metal type of artefacts used *within* the mines with those of metal produced *from* the mines has been encountered on the Great Orme, where fragments of bronze believed to have been detached from the end of a chisel(s) were subjected to analysis, following which some very different interpretations were given as to their age and source (Dutton & Fasham 1994,279; Rohl & Needham 1998, 111; Northover 2002).

As suspected, a study of Early Bronze Age metalwork from Cardiganshire, including from the vicinity of the mines themselves, tells us little about mining and metal production during the period in question. Briggs (1994, 155) lists four flat axes of copper or bronze, a halberd, and several socketed spearheads, whilst another copper axe from Tirmynach is listed by Savory (1980). None of these appear to have been found close to the sites of prehistoric mining, indeed the majority of the axes appear to have been earlier, and probably of 'Irish type' metal. Interestingly, the halberd from Pontrydygroes was found 'near an old copper mine' (Logaulas Mine), but again this was made from Irish metal, consistent with a Kerry source (Northover 1980). Nevertheless, Briggs (1994, 157) suggests that both this and the Rhydypennau flat axe were conceivably of local metal and workmanship, although he admits that no casting moulds or other evidence for local metalworking of this period had yet been found anywhere within the area. Furthermore, whilst there are broadly similar numbers of Middle and Late Bronze Age metalwork finds from Ceredigion, few of the Earlier Bronze Age examples appear to date from the actual mining period itself (1900-1700 BC).

The use and availability of resources and the impact of mining on the environment

Given the fairly comprehensive local palaeo-environmental record constructed for the mining period we might well expect to see some useful correlations emerge between intermittent short-lived declines in arboreal pollen (particularly of oak) and the exploitation of woodland as fuel for firesetting, but possibly also for mine timber and charcoal for smelting. Selective use of oak for the purposes of firesetting has also been noted at Mt.Gabriel (McKeown in O'Brien 1994; Mighall et al. 2000 d), and one might compare this with other preferred uses of this as a fuel for specific activities such as in cremation pyres (Lynch et al. 2000, 81). On Copa Hill the coppicing of hazel probably had much less of an environmental impact, although it is possible that the timing of this may also have coincided with, or post-dated a more general tree clearance which occurred during the earlier Bronze Age, a pattern noted elsewhere within the Plynlimon hinterland and thought to be associated with transhumance agriculture, the renewed occupation of the uplands, and perhaps even prospection for metals (Moore 1968, 1009; Timberlake 2002). As has already been suggested, the repetitive small scale declines in tree cover witnessed on Copa Hill could have been associated with renewed phases of localised mining activity carried out over a three to four hundred year period, a similar pattern to that noted at other prehistoric mining and early iron working sites such as Mt Gabriel (Mighall et al. 2000 c) and Bryn y Castell hillfort (Mighall and Chambers 1993). Indeed, much of the palaeoecological data obtained from peat bogs close to prehistoric mines and metalworking sites suggests that the impact of mining on local woodland in prehistory was often negligible and any local deforestation short-lived (Dorfler 1995; Marshall 1999; Mighall & Chambers 1993b; 1997; Mighall et al. 2000 b; Pott et al. 1992).

Independently of the pollen evidence from Copa Hill, some calculations of the consumption of wood in firesetting and timbering are possible based on the known tonnage of rock extracted, and from an understanding of the ratio of wood: rock required as a result of firesetting experiments (Lewis 1990 b Crew 1990; Timberlake 1990 c; O'Brien 1994). O'Brien (1994, 169) found that the ratio of unburnt wood to rock (weight) removed by stone hammers/chisels following firesetting experiments on Mt.Gabriel was much lower (1:0.05 and 1:0.27) than that recorded by Lewis (1990 b) upon the Great Orme (1:1.08 and 1:1.5), or Timberlake (1990) at Cwmystwyth (1:2.3). However, more recent experiments carried out at Penguelan on Copa Hill (unpublished) appear once again to confirm that the likely ratio here was about 1:2 (wood fuel:rock), or better still. Thus the extraction of about 5000 tons of rock may have required up to 2500 tons of wood, therefore 3 to 4 thousand mature trees over a period of 400 years, or as few as 5 trees per year (compare this with an estimate for the consumption of 2-15 K tons of wood over a putative 200 year period at Mt.Gabriel (O'Brien 1994, 291)). Whilst such ideas of 'total constancy' in the rate of working are unrealistic, periodic mining campaigns involving the breaking of 5-10 cubic metres of new ground may have occurred every couple of years. At present the evidence here for smelting is too weak to include within the calculations for the consumption of wood, such as in the manufacture of charcoal. As regards still other utilised materials, the evidence for wastage, replenishment and re-use of cobbles as mining hammers implies that upwards of 5-10 thousand of these stones may have been brought up to site during the period of operation of the mine.

It could be argued that environmental change in the form of partial de-afforestation and the development of blanket peat and open moorland above the mine during the Early Bronze Age was brought about or hastened by mining (Mighall & Chambers 1993,264). We do at least now have some idea of the scale and type of local woodland cover present prior to the commencement of mining upon the hill, such as the pollen data from the base of the peat monolith CH3 which suggests an arboreal pollen percentage of between 44 and 56% TLP prior to 2600 BC (Mighall et al. 2002 c), whilst the continued dominance of oak and hazel woodland through the mining period, and immediately after it seems to suggests that woodland depletion for mine fuel could not itself have been sufficient to bring about this fundamental change of habitat. The pollen record both during and after this time also seems to imply other human activity apart from mining, notably the presence of grazing animals (also suggested by the beetle evidence) and very minor cereal cultivation, yet a more far-reaching influence such as climate deterioration might also be invoked. However, the latter agent as a cause for blanket peat formation is considerably more speculative. Instead it seems much more likely in this case that gradual woodland clearance on the plateau was responsible. This accelerating effect may have been much more noticeable within areas which had already suffered woodland clearance and where the same degree of natural regeneration subsequently proved impossible. Lynch et al (2000, 83) suggest that an increase in pastoral farming and in particular sheep grazing within upland Wales would have been one of the most effective brakes to regeneration from the Early Bronze Age onwards. As elsewhere in Wales (Caseldine 1990, 55-66) this process of peat formation taking place on Copa Hill seems to have been already underway at the beginning of the Bronze Age, yet recent evidence also implies the higher parts of the plateau may never have been forested at all. Nevertheless, peat encroachment could have meant, just a few hundred years after mining commenced, that the days of this operation were numbered; the mine beginning to flood due to the slow but steady inflow of water issuing from the area of peat mire which lay above it. Another factor which could have contributed to its demise, apart from flooding and the near exhaustion of rich surface ore, may have been de-population within the upland zone. This had begun to affect the Upper Ystwyth Valley and Plynlimon hinterland before and after 1500 BC (Lynch et al. 2000). Thereafter, the drift of people away from the uplands would have affected mining as well as agriculture, and in particular the incidence of settlement within the area.

During 1991 the discovery of small amounts of fossil wood (mostly birch *Betula* rootlets) within the sides of erosion gullies dissecting the blanket peat on the top of Copa Hill (Briggs 1991; Timberlake & Mighall 1992), began to fuel an existing debate over whether this had been the source of the wood used by the earliest miners, thereby providing an erroneously early date for the charcoal collected from the mine, and thus by inference the validity of Bronze Age mining in Wales as a whole (Briggs 1988 & 1991; Budd et al. 1992). Since then the consistency and reproducibility of C 14 dates obtained from both wood, charcoal, antler and bone sampled from at least 8 other excavated early mining sites seems to have ended most speculation on this matter. However, there are other interesting points here which relate to the use and existence of fossil fuel sources as well as to claims that sufficient copper ore could be recovered from superficial deposits (on Copa Hill) without recourse to mining (Briggs 1991). For example, the fossil wood horizon on Copa Hill was fairly conclusively shown to be later than the Bronze Age dates obtained for mining (Timberlake & Mighall 1992, 41), as well as being unrepresentative of the actual fuel used at the mine (mostly immature cut oak). The size of this resource was also quite inadequate to account for the amount of firesetting which had actually taken place. Larger amounts of fossil wood may be present within the peat at Borth Bog close to the site of the prehistoric mine at Llancynfelin, yet there are question marks over its combustibility (Timberlake & Mighall 1992, 43), or indeed the general suitability of using peat or peat wood fuel for firesetting (Timberlake 1990c). Nevertheless, peat was used for this purpose during the medieval period in the Weardale mines (Blackburn 1994). Similarly, the collection of small amounts of oxidised ore from scree located below the lode outcrop on Copa Hill is a distinct possibility, yet Briggs' suggestion of much larger amounts of superficial ore covering the slopes (1991, 5) seems doubtful, since it is quite evident that the majority of the hillside and its mineral veins were still concealed beneath a thick mantle of glacial moraine. Moreover, the concept of Bronze Age man scavenging sufficient amounts of copper from widely dispersed glacially transported boulders is much more difficult to grasp than the possibility that he first collected and used alluvial, eluvial or colluvial ore which would have lain much closer to the primary mineral source. Indeed it has already been suggested that this may have been the principal method by which Bronze Age prospectors first discovered these veins in mid-Wales, the presence of residual 'shoad lead ore' providing them with the main clues for prospection (Morris 1744 in Bick & Wyn Davies 1994; Timberlake 2002a), whereas chalcopyrite was much more likely to oxidize and disintegrate after being transported any appreciable distance, thus rarely if ever surviving in amounts sufficiently large enough to work (Timberlake & Mighall 1992, 43).

The social organization and significance of mining in the context of use and occupation of the upland zone

In the absence of any evidence for an associated mining camp, valley settlement, or ceramic /metalwork culture, one might expect an interpretation of the social organization, economics, or seasonal cycle of mining to be highly speculative, and of therefore of limited value. However, a good deal can still be inferred from both tools, chronology, and the form of the workings. Thus the use of children as miners on the Great Orme has been suggested on the basis of the discovery of extremely tight passages underground (Lewis 1994, 36), whilst a model of continuous and relatively rapid (rather than spasmodic) exploitation of individual mines upon Mt.Gabriel was suggested as an explanation of how flooding, and thus the de-watering of the inclined adits was avoided (O'Brien 1994). On Copa Hill, the presence of regularly replenished beach pebbles, some of which show considerable use, and others which show hardly any use at all, is suggestive of intermittent supply and seasonal work campaigns, with the miners coming from afar, a parallel that Paul Craddock (1994, 74; 1995, 107) draws uopn with the working of the large iron meteorites in the Cape York region of Greenland by the Inuit over many hundreds of years using hammer-stones of basalt, some carried for distances of more than 50 kilometres from source. On this basis, as well as the evidence for fuel wood and foliage cut and brought to the mine during the late Spring/early Summer months, it has been suggested that the miners on Copa Hill probably came from the coast, travelling up the Ystwyth Valley, bringing with them fresh tools and sufficient resources to last the summer season. Such part-time activity could thus be linked to transhumance agriculture and pastoralism, as well as with family groups – the latter bringing their flocks of sheep, goats, or cattle to graze on summer pasture within the uplands, whilst at the same time attending to the diggings at the mine. There are modern analogies to this in the form of squatter mining as well as the mining of mineral specimens by both shepherds and mountain guides within the Himalayan region and the Atlas mountains of North Africa (Timberlake 2002a, 184).

The work of mining would not necessarily involve all the community, perhaps only a portion of the male members (perhaps as few as ten in any one season), individuals who would in any case be provided for by the nearby the agricultural camp. More interestingly, one indirect result of the clearance of new land for pasture and the subsequent erosion caused by de-afforestation could have been the discovery of mineral veins, and thus the claim to work these as mines. An explanation perhaps of why the prospection for metals was the province of agriculturists rather than of specialist metallurgists, the right to work them being inherited from generation to generation thereafter. Such a scenario may well have been typical of all or most of these smaller upland mines, a fact which could help to explain the obvious lack of specialisation, any form of permanent settlement, or perhaps even the absence of smelting, particularly if the hand-picked ore concentrate was traded further afield with metallurgists or else their intermediaries, and exchanged for prestige goods. It is perhaps useful here to make a comparison with Shennan's socio-economic model for copper production within the Bronze Age mining settlement of Klingleberg in the Mitterberg, Austria (Shennan 1999). The latter was evidently part of a much bigger production centre where ore was smelted from the hinterland mines, yet there appeared to be no evidence here for elite consumption or even high status occupation equivalent to that of the Late Bronze Age salt miners of Hallein, and thus he concluded that the mining region of the Tyrol was a marginal zone, with miners and smelters producing copper for relatively low return simply as a means to enter the exchange economy. Perhaps there is a salient point here. By the middle of the Early Bronze Age the level of production in Britain may have been such that copper ore or even copper metal may not have been worth much at source. Thus mining may well have been the prerogative of a people who have left us little in the way of permanent monuments (such as burial cairns) or a ceramic culture, but of whom evidence may be found elsewhere within the landscape in the form of burnt stone mounds, features common within the uplands of Britain, particularly to West Wales (Caseldine & Murphy 1989).

It has been suggested that burnt stone mounds or charcoal filled pits (many of which date to the earlier part of the Bronze Age), as well as being cooking places, sweat lodges or the sites of fires associated with agricultural clearance, might include the evidence of primitive industrial activity such as smelting or metalworking (Barfield & Hodder 1987), a theory supported by the finding of numerous burnt pits and charcoal spreads near Pentraeth and Llantrisant in Anglesey, within one of which was found a small fragment of slag (Lynch 1991; White 1977). Within Ceredigion there appears to be no association of burnt mounds and Bronze Age mines, yet the distribution of these mines is still interesting, in that there appears to be some evidence for a grouping of sites and an association with east-west valley routes into the Cambrian Mountains. In one case this is with a possible trackway or territorial boundary that crosses the mountain divide, skirting around the base of Plynlimon, a route which in some places is marked by an alignment of stones (Timberlake 1994; Timberlake 2002a; SEE Fig.102 *this vol.*). The latter extends eastwards from the area of the Beaker-Early Bronze Age ritual complex near Penrhyncoch (Houlder 1957; Murphy 1986) along a ridge route that passes within a few yards of the important but putative prehistoric mine of Twll y mwyn - Darren. East of Plynlimon the same route can be picked up where it descends into the tributary valleys of the River Severn, past the mines of Nantyreira and Nantyrickets. The area of active mining and prospecting in all cases appear to be within the marginal zone surrounding the ritual cairn-filled landscape of Plynlimon. A more lowland setting has also been identified which includes several prehistoric copper mining and prospection sites that surround Borth Bog and the Dovey Estuary (Timberlake 1995- AW; 1997; Timberlake 2002b).

Whilst such associations between mines and other elements in the prehistoric landscape are invariably speculative, such evidence can be used in support of the argument that these were not necessarily so remote or marginal after all, and indeed were linked by the very same networks used for long-distance trade between Ireland and England.

Figure 102: Early Bronze Age field monuments and mines within the Plynlimon area

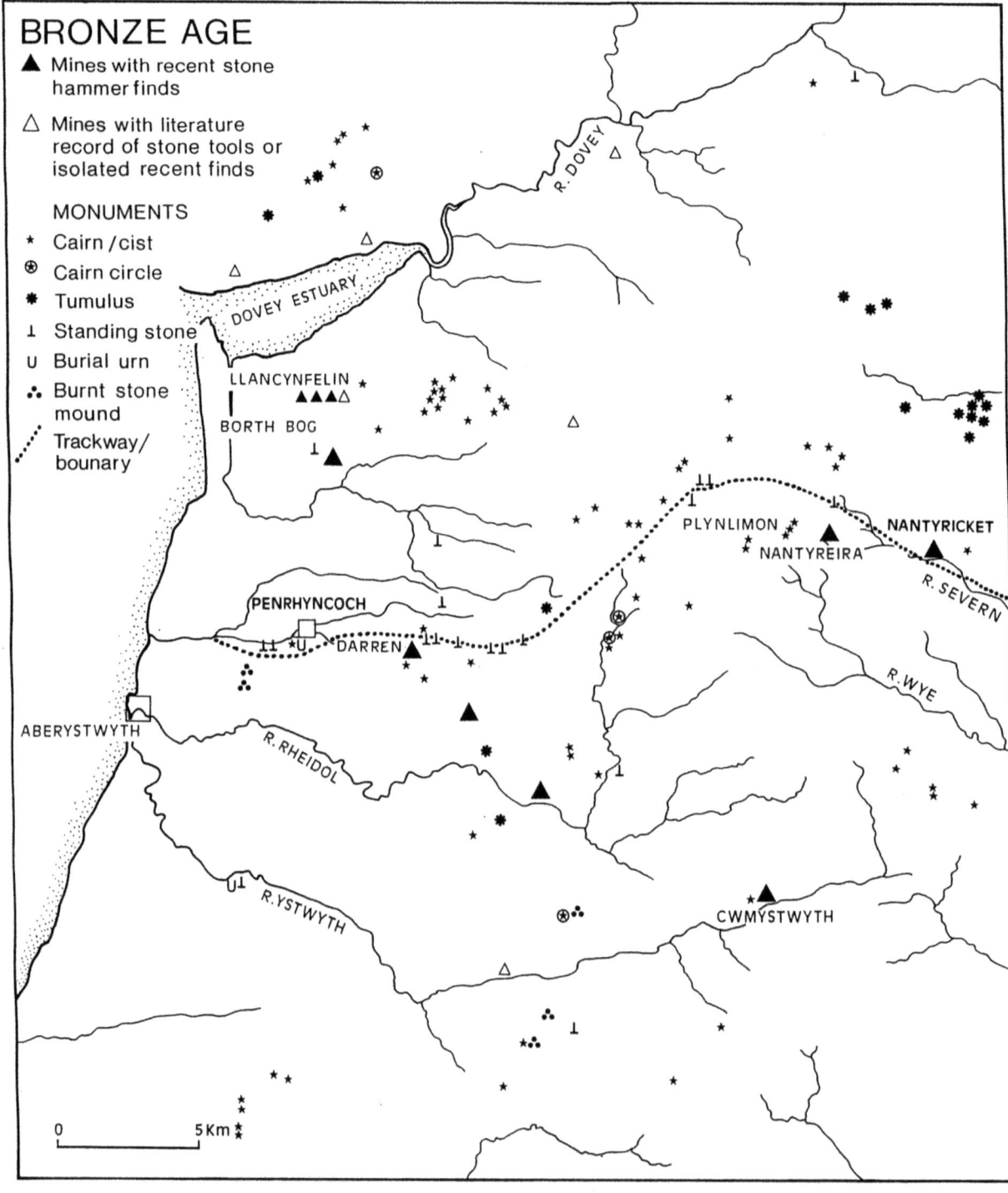

CHAPTER 19

SUMMARY CONCLUSIONS

There seems little doubt now that both Britain and Ireland experienced a widespread phase of prospection and small-scale mining for copper and other ores at the very beginning of the Bronze Age (2000 - 1700 BC) following the development of more localised metal working industries and the exploitation of a multiplicity of different ore sources accompanying the first use of indigenous tin bronze. For reasons as yet unknown Wales appears to have been one of the main regions where such mining was carried out, with this area of West Wales (Ceredigion and West Montgomeryshire) containing the highest number of known prospection sites. Of these, Copa Hill is the best understood and most thoroughly investigated example, and is perhaps typical of a small-medium sized opencast trench mine which exploited impersistent but locally rich pockets of chalcopyrite present within copper-lead-zinc sulphide veins cutting these Lower Palaeozoic rocks. Metal produced from this ore would have been free of significant impurities, although very minor traces of lead, nickel, cobalt and arsenic may be characteristic of the unmixed copper, and it is possible that metal from this or similar sources contributed to recognisable types of Early Bronze Age metalwork (e.g. Willerby etc.).

The earliest date for prospection activity on Copa Hill presents us with something of a dilemma, since here there appears to be traces of Mesolithic activity, as well as later Copper Age (c.2500 BC) extraction of ore located upon the actual site of the exposed vein, the latter associated with very small waste tips from processing. However, more substantial opencast mining carried out from the downslope end of the trench was clearly underway by 2000 BC, the mine reaching its maximum depth about a hundred years later, by which time it is evident that the ingress of water into the workings was already a serious problem. Yet the mine evidently continued to operate after the deepest sections had been abandoned, the main vein being worked northwards (up-slope) on a more piecemeal basis over the next two to three hundred years. It was this waterlogging and the continued infill of the workings with peat and washed -in mine spoil following its final abandonment around 1600 BC which indirectly led to the preservation of wooden artefacts, including the wooden drainage launder found lying within the entrance cutting to the mine, and recognised as being one of the most perfectly preserved and almost certainly the oldest example of such mine equipment known.

The collection and transport to site of thousands of waterworn cobbles for use as mining hammers, crushing stones and anvils, many of them showing signs of substantial use and re-use following breakage, is perhaps one of the most characteristic and enduring features of these primitive prehistoric mines. Yet apart from the evidence for firesetting and mine spoil, few other identifiable remains such as those of habitation or smelting were found anywhere within the vicinity of the opencast, thus this working is rather typical of the nine primitive mines already investigated which have produced no ceramic or metalwork remains, similar in some respects therefore to those sites of burnt stone mounds common throughout the upland landscape of Britain (and perhaps also to the utilitarian culture which produced them). Indeed, the recognition of many of the broken stone tools as beach cobbles brought up the Ystwyth Valley to the site from the coast some 25 km distant supports other archaeological evidence from the mine for the existence of seasonal working, thus it is suggested that the miners may have been pastoralists who brought their animals inland to graze on upland pastures during the summer months. Thus despite the large volume of rock and ore removed (5000 metric tonnes) the annual rate of working and physical demands upon the environment in terms of the numbers of trees felled for fuel etc. may really have been quite small in terms of the timescale over which this was worked - a deduction supported by the palaeo-environmental evidence from the site. Thus the mine may have produced as little as 1-2 tons of copper in total, and it would seem feasible therefore that this could have been traded by the miners piecemeal as ore concentrate rather than as smelted and refined metal.

Ten years of excavations on Copa Hill have provided us with a fairly good understanding of the types of ore sought and also the techniques employed in prospection and mining at these upland mines from the beginning of the Bronze Age onwards. However, logistical problems of archaeological excavation at this isolated site prohibited the complete sectioning of the infilled opencast and thus the reaching of mining sediments at the very base of the working, within which further well preserved wooden artefacts may have lain. Furthermore, the chronological complexity of ancient mine workings, only really resolvable through careful radiocarbon dating of tip sequences, would have justified the inclusion of a much finer resolution programme of sampling if resources here had been available. The absence of evidence for smelting, either at the mine or elsewhere within the surrounding archaeological landscape, also leaves unanswered the question as to exactly what type of metal was being produced, as well as the method and efficiency of the process. Moreover, it seems rather doubtful whether an increased sampling strategy of the mining hinterland would have been any more likely to reveal signs of smelting or evidence for the ephemeral mining camps such as have been found on Mt.Gabriel and Ross Island (O'Brien 1994; 1996), but which appear to remain undetected within Britain. Meanwhile, the apparent extraction of lead minerals raises interesting and controversial questions as to the use of this metal, or experimentation with it, at the very beginning of the Bronze Age, and as such this is something that both archaeologists and archaeo-metallurgists should be made aware of.

Contrary to what was once thought, much of the earliest evidence for the extraction of metal ores seems to have survived the depredations of later mining, thus it seems likely that other sites still remain to be discovered elsewhere within Britain. The resulting body of steadily gathering data has and will continue to be invaluable in helping to understand such questions as the provenance and changing provenance of metal used, the amount of metal in circulation, perhaps even its contribution to local economies and role in local and long distance exchange networks within early Bronze Age Britain.

The existence as early as 2000 BC of an indigenous metalworking industry independent of supplies from Ireland and the continent, and using Welsh or other British copper plus tin from south-west England, now seems to be a fairly well substantiated reality, rather than an assertion based on tradition and conjecture.

Figure 103: Cross-sectional profile through centre-front of Comet Lode Opencast showing prehistoric-historic mining and abandonment horizons (and features). Based on stratigraphic and absolute dating evidence. Drawing by B.Craddock from info. supplied by ST, revised 2001

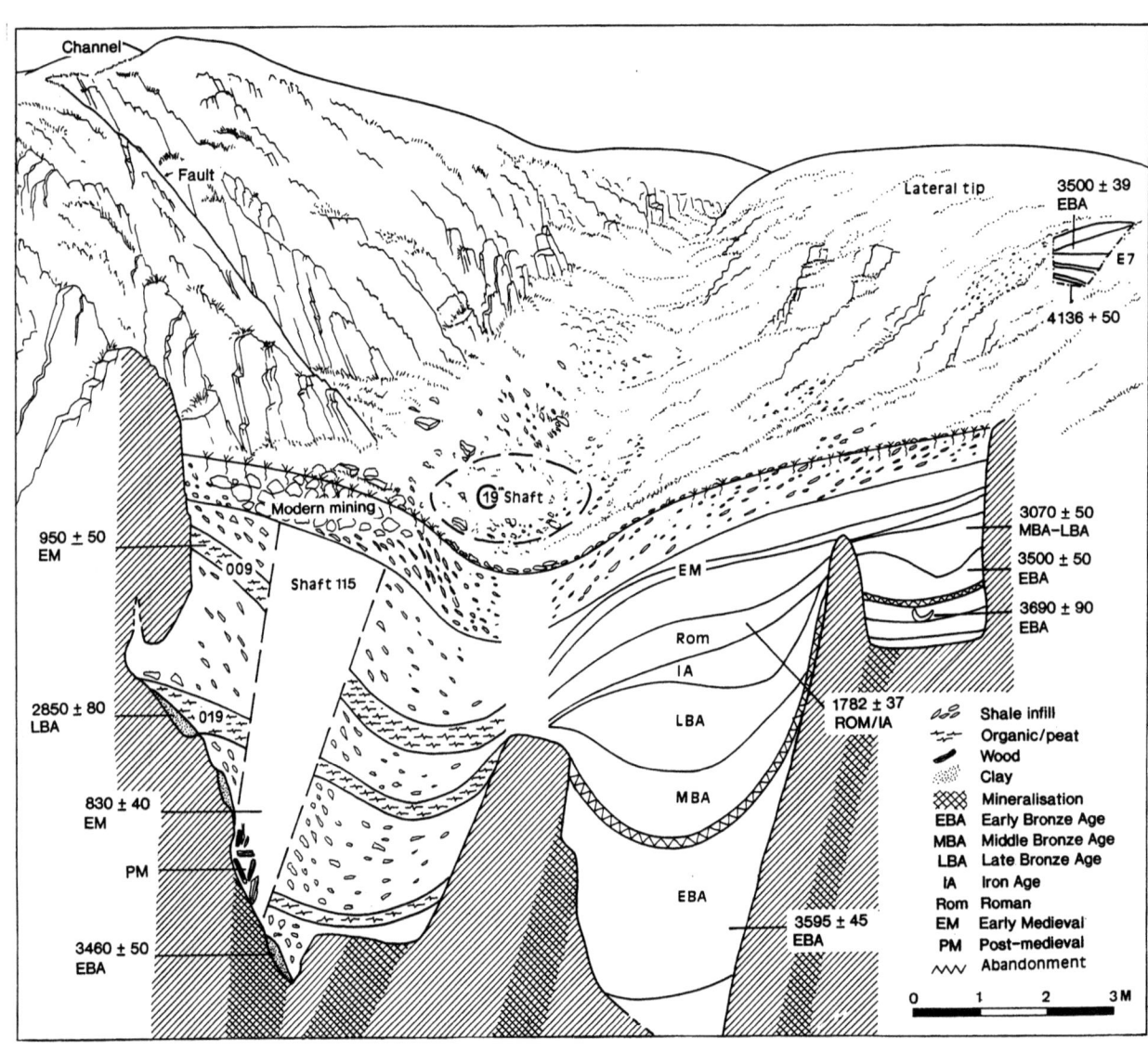

Acknowledgements.

The completion of this project owes much to the dedication and hard work of the experienced but volunteer based membership of the Early Mines Research Group, in particular Brenda Craddock, Phil Andrews, Anthony Gilmour, Jamie Thorburn, Tom Doidge and numerous other excavators and helpers between the years 1989-1999, whilst Duncan James assisted me in the very first trial dig of 1986. Furthermore, I owe a particular debt of gratitude to those archaeologists who encouraged and supported me in this work during its early stages, in particular the late Douglas Hague, the late George Boon, Frances Lynch, Peter Crew, David Jenkins, Paul Craddock, Stephen Aldhouse-Green and Ken Brasil. Meanwhile long term support for the project which included financial assistance offered by the National Museum and Galleries of Wales, Cardiff enabled regular seasons of excavation to continue thereafter. Thus I would also like to acknowledge the help of Richard Bevins, Richard Brewer and Adam Gwilt in recent years. RCAHM(Wales) and Dyfed Archaeological Trust (Cambria Archaeology) provided assistance with survey work and SMR data. The Department of Scientific Research at the British Museum kindly offered to fund some of the radiocarbon dating as part of its own research programme, whilst a similar series of dates was also awarded under the NERC Accelerator Dating Service at Oxford. The original series of dates in 1986 were provided free of charge by the Godwin Laboratories, Cambridge courtesy of Roy Switsur. In addition to that provided by NMGW, grant aid has also been awarded on several different occasions by the British Academy, Society of Antiquaries, Royal Archaeological Institute, Prehistoric Society, Cambrian Archaeological Association and the Historical Metallurgy Society, to all of whom I am grateful. Permission to excavate (SMC) was granted by CADW and by the landowners the Crown Estate, with the additional agreement of the tenant farmer Mr.J.Raw, Ty Llwyd. The Council for British Archaeology provided a small publication grant to assist with the costs of producing the colour plates at the front of this monograph.

BIBLIOGRAPHY

Adams, D.R. 1970 *The Mines of Llanymynech Hill* (updated A.J.Preece 1992), Shropshie Caving & Mining Club Account No.14

Agricola, G. 1556 *De Re Metallica*, trans. H.C. & L.H.Hoover (1950), Dover Publications, New York, 638 pp

Ambers, J and Bowman, S. 1998 Radiocarbon determinations from the British Museum: Datelist XXIV, *Archaeometry 40 (2)*, 413-435

Ambert, P., Barge, H.., Esperou, J.L. & Bourhis, J.R.1983Mines de cuivre prehistoriques de Cabrieres (Herault) premiers resultats IN Coddet, C. (ed.) *Journees de Paleometallurgie*, Compiegne, 225-236

Andrews, P. 1994 Excavating Mines (Copa Hill, Chinflon, Kestel) IN Ford,T.D. & Williesm, L. (eds.) 'Mining Before Powder' *Bull. Peak District Mines Hist. Soc. Journ. 12 (3)*, 13-21

Anon 1859 Catalogue of the contents of the museum (Report of the Cardigan Meeting). *Archaeologia Cambrensis, new series Vol.1*, 331-334

1861 Catalogue of the contents of the temporary museum in the ancient hall of Swansea Castle, during the meeting at Swansea in 1861. *Archaeologia Cambrensis, 3rd.series 7*, 367-372

1866 Temporary Museum at Machynlleth (Report on the 20th annual meeting of the Cambrian Archaeological Association). *Archaeologia Cambrensis, 3rd. series 12*, 544-549

Armfield, C. 1989 Dressing floors on the Kingside Lode, Copa Hill, Cwmystwyth, *Archaeology in Wales 29*, 27-29

Arnal, J.A. & Bocquet, A., Robert, A. & Verraes, G. 1979 La naissance de la metallurgie dans le sud-est de la France, IN Ryan, M. (ed.) *The Origins of Metallurgy in Atlantic Europe*, 1980, Dublin

Barber, K.E. 1976 History of vegetation. In (S.B. Chapman, Ed.) *Methods in Plant Ecology*. Oxford: Blackwell, pp5-83.

Barfield, L. & Hodder, M. 1981 Birmingham's Bronze Age, *Current Archaeology 78*, 198-200

Barnatt, J. 1999 Prehistoric and Roman mining in the Peak District: present knowledge and future research, *Mining History: The Bull. Peak Dist. Mines Hist. Soc. 14 (2)*, 19-30

Barnatt, J. & Thomas, G. 1998 Prehistoric mining at Ecton, Staffordshire: a dated antler tool and its context, *Mining History: Bull. Peak Dist. Mines Hist. Soc., 13 (5)*, 72-78

Barnes, J.W. 1979 The first metal workings and their geological setting IN Crawford, H. (ed.), 1979, 44-84

Bartelheim, M., Pernicka, E. & Krause, R. (eds.) 2002 Die Anfange der Metallurgie in der Alter Welt/ The Beginnings of Metallurgy in the Old World, *Archaometrie – Freiberger Forschungen zur Altertumswissenschaft 1* (Proceedings of Conference held at TU Bergakademie Freiberg, Germany in November 1999)

Bell, M. 1993 Field Survey and Excavation at Goldcliff, Gwent 1993, IN *Archaeology in the Severn Estuary 1993 - Annual Report of the Severn Estuary Levels Research Committee*, M. Bell (ed.), Lampeter, 81-102

Bevins, R.E. & Mason 1997 Welsh metallophyte and metallogenic evaluation project. Results of a mine site survey of Dyfed and Powys, CCW Contract Science Report No. 156 (National Gallery and Museums of Wales)

Bick, D. 1976 *The Old Metal Mines of Mid-Wales Part 3 Cardiganshire - north of Goginan*, The Pound House, Newent, Glos., 72 pp

1991 *The Old Metal Mines of Mid-Wales Part 6: A Miscellany*, The Pound House, Newent, Glos.

1995 Note: Hematite mines in Swaziland, *Historical Metallurgy Society Newsletter no.29*, Spring 1995

1999 Bronze Age copper mining in mid Wales - fact or fantasy?, *The Journal of Historical Metallurgy 33, No.1*, 7-12

Bick, D.E. & Davies, W., 1991 *Lewis Morris and the Cardiganshire Mines*, National Library of Wales, Aberystwyth, 89pp

Bird, J.B. 1979 The 'Copper Man': A Prehistoric Miner and his Tools from Northern Chile. IN *Pre-Columbian Metallurgy of South America* ed. E.P. Benson, Dunbarton Oaks, Washington, 359-371

Blackburn, A. 1994Mining without Laws: Weardale under the Moormasters, IN Ford, T.D. & Willies, L. (eds.) *Mining Before Powder*, 69-75

Bogosavljevic, V. 1988 Materiel archeologique meuble de la mine prehistorique de Prljusa-Mali Sturac IN Jovanovic, B.: *Recherches sur l'Exploitation Miniere et la Metallugie Anciennes dans la Region du Radnik*. Cacak.

Briggs, C.S. 1983 Copper mining at Mt.Gabriel, County Cork: Bronze Age bonanza or post-famine fiasco?, *Proceedings of the Prehistoric Society 49*, 317-333

1988 The location and recognition of metal ores in pre-Roman and Roman Britain and their contemporary exploitation IN Ellis-Jones, J. (ed.) 1988, 106-114

1991 Early Mining in Wales: the date of Copa Hill, *Archaeology in Wales 31*, pp.5-7

1993 Early Mines in Wales again: a reply to Budd et.al 1992, *Archaeology in Wales 34*, 13-15

1994 The Bronze Age: Chapter 2 IN Davies & Kirby (eds.) 1994, 124-218

Bromehead, C.E.N. 1954 Mining and Quarrying IN *A History of Technology Vol. 1* (ed. Singer,C., Holmyard, E.J., Hall, A.R. and Williams, T.I.), Clarendon Press, Oxford, 558-571

Bronk Ramsey, C. 1995 Radiocarbon Calibration and Analysis of Stratigraphy: The OxCal Program, *Radiocarbon 37*, 425-430

2000 Oxcal computer program (v3.5), Proc. 15th Internatl. C14 Conf., *Radiocarbon 37 (2)*, 425-430

Bruce, J.L. 1937 Antiquities in the Mines of Cyprus, Appendix V, *The Swedish Cyprus Expedition Vol. 3*, 639-671

Budd, P. 1993 Recasting the Bronze Age, *New Scientist* No.1896. Oct. 1993, 33-37

Budd, P. & Gale, D. 1994 Archaeological survey of an early mine working at Wheal Coates, near St.Agnes, Cornwall, *Cornish Archaeology 33*, 14-21

Budd, P. & Gale, D. (eds.) 1997 *Prehistoric Extractive Metallurgy in Cornwall*, Proceedings of the Camborne Conference July 1992, 57 pp.

Budd, P., Gale, D., Pollard, A.M., Thomas, R.G., & Williams, P.A. 1992 a The early development of metallurgy in the British Isles, *Antiquity 66 (252)*, 677-686

1992 b Early Mines in Wales: A Reconsideration, *Archaeology in Wales 32*, 36-37

Burnham, B. 1994 Dolaucothi Revisited IN *Mining Before Powder*, eds. Ford & Willies, 41-47

1997 Roman Mining at Dolaucothi: the Implications of the 1991-93 Excavations near the Carreg Pumpsaint, *Brittania XXVIII*, 325-336

Burnham, B. & H. 1990 Annel & Gwenlais Leats, *Archaeology in Wales 30*, 55

Burnham, B & H. & Walker, M.J.C. 1992 Excavations across the Annell & Gwenlais Leats near Dolaucothi in 1990, *Archaeology in Wales 32*, 2-8

Caseldine, A.E. & Murphy, K. 1989 A Bronze Age burnt mound on Troedrhiwgwinau Farm near Aberystwyth, Dyfed, *Archaeology in Wales 29*, 1-5

Cauuet, B. (ed.) 1999 l'Or en Antiquite, de la Mine a l'Objet, *Aquitania, suppl.IX*

Cauuet, B., Ancel, B. & Cowburn, I. 2000 *The Dolaucothi Gold Mines: The Intermediate Report (August 2000)* of the Archaeological Appraisal by a Franco-British Team (April 2000) for The National Trust, 75 pp + plans (unpublished)

Cernych, E.N. 1978 Aibunar - A Balkan copper mine of the fourth millenium BC, *Proceedings of the Prehistoric Society 44*, 203-218

Chapman, D. 1997 Great Orme Smelting Site, Llandudno, *Archaeology in Wales 37*, 56-57

Claris, P. & Quatermaine, J. 1989 The Neolithic Quarries and Axe-Factory sites of Great Langdale and Scafell Pike, *Proceedings Prehistoric Soc. 55*, 1-26

Clark, P. 1997 Lessons from Bronze Age boat-building, *British Archaeology No.24*, May 1997, p.7

Clark, S.H.E. 1997 *Copper and Coleoptera - fossil coleopteran evidence for the impact on the environment of prehistoric mining at Copa Hill, Cwmystwyth, Wales*, MSc. dissertation in Environmental Archaeology and Palaeoeconomy, University of Sheffield 1996-7, unpubl. 32 pp

Claughton, P. 1996 The Lumburn Leat - evidence for new pumping technology at Bere Ferrers in the 15th century, in Newman, P. (ed.) *Mining History 13*, 35-40

Coghlan, H.H. & Case, H.J. 1957 Early metallurgy of copper in Ireland and Britain, *Proc. Prehistoric Soc. 23*, 91-123

Condry, W.1994 The vertebrate animals of Cardiganshire, Chapter 5 IN Davies, J.L. & Kirby, D.P. (eds.) *Cardiganshire County History*

Coope, G.R. & Osborne, P.J., 1968 Report on the Coleopterous fauna of the Roman well at Barnsley Park, Gloucestershire. *Transactions of the Bristol & Gloucestershire Archaeological Society 86*, 84-7.

Craddock, B.R. 1990 The experimental hafting of stone hammers, IN Crew & Crew (eds.) *Early Mining in the British Isles* (1990), 58

1994a Drawing ancient mines, IN Ford, T.D. & Willies, L. (eds.) *Mining Before Powder* (1994), 9-12

1994 b Notes on stone hammers In Ford & Willies (eds.), 28-30

Craddock, P.T. 1980 Scientific Studies in Early Mining and Extractive Metallurgy, *British Museum Occasional Paper No.20*, Craddock, P. (ed.) BM London, 173 pp

1994 Recent progress in the study of early mining and metallurgy in the British Isles, *Journ. Historical Metallurgy Soc. Vol.28 (2)*, 69-85

1995 *Early Metal Mining and Production*, Edinburgh University Press, 363 pp.

Craddock, B.R. & Craddock, P.T. 1996 The beginnings of metallurgy in south-west Britain: hypotheses and evidence, in *The Archaeology of Mining and Metallurgy in South-West Britain, Mining History: Bulletin PDMHS, Vol.13 (2)*, ed. Philip Newman, 52-62

Craddock, P.T. & Craddock, B.R. 1997 The inception of metallurgy in South-West Britain: hypotheses and evidence, IN Budd & Gale (eds.), 1997, 1-14

Crawford, H. 1979 *Subterranean Britain - Aspects of Underground Archaeology*, John Baker, London, 201 pp.

Crew, P. 1990 Firesetting experiments at Rhiw Goch, 1989, IN Crew, P. & Crew, S. (eds.) *Early Mining in the British Isles*, 57

Crew,P. & Crew,S. (eds.) 1990 *Early Mining in the British Isles*, publ. Plas Tan y Bwlch Snowdonia National Park Study Centre, Maentwrog, 80 pp.

Cummins, W.A. 1979 Neolithic stone axes:distribution and trade in England and Wales, IN McK Clough, T.H. & Cummins, W.A. (eds.), Stone Axe Studies, *Council for British Archaeology Research Report 23*, London, 5-12

Cwrt Mawr deeds - Mss in National Library of Wales (Cwrt Mawr 208 etc.)

Dahm, C., Lobbedey, U. & Weigerber, G. 1998 *Der Altenberg -Bergwerk und Siedlung aus dem 13. Jahrhundert im Siegerland*, Denkmalpflege und Forschung in Westfalen 34 (ed. R. Habelt, Bonn), 263 pp

David, G.C. 1996 Great Orme Bronze Age Mine, *Archaeology in Wales 36*, 59-60

1997 Great Orme Bronze Age Mine, *Archaeology in Wales 37*, 56

1998 Great Orme Bronze Age Mine, *Archaeology in Wales 38*, 94-95

2000 Great Orme Bronze Age Mine, *Archaeology in Wales 40*, 73-75

Davies, J.L. & Kirby, D.P. (eds.) 1994 *Cardiganshire County History Volume I: From the earliest times tothe coming of the Normans*, publ. on behalf of Cardiganshire Antiquarian Soc. in assoc. with RCAHM (Wales), University of Wales Press, Cardiff

Davies, J.R. 1997 Geology of the country around Llanilar and Rhayader: Memoir of the British Geological Survey Sheet 178 (England & Wales)

Davies, O. 1935a *Roman Mines in Europe* Macmillan, London

1935b Mss. notes on mines of mid and North Wales, RCAHM(Wales)

1936 *Report of Annual Meeting of the British Association for the Advancement of Science, Section H: Early Mining Sites in Wales*, 304-305

1937 *Report of Annual Meeting of the British Association for the Advancement of Science, Section H 4: Mining Sites in Wales*, 301-303

1938a *Report of Annual Meeting of the British Association for the Advancement of Science, Section H: Mining Sites in Wales*, 342-343

1938b Ancient Mines in Montgomeryshire, *Montgomeryshire Collections 45*, 55-60 & 152-157

1939 Excavations on Parys Mountain, *Transactions of the Anglesey Antiquarian Society and Field Club*, 40-42

1939/1940 *Report of Annual Meeting of the British Association for the Advancement of Science: Early Mining Sites in Wales*, 126-127

1947 Cwm Ystwyth Mines, *Archaeologia Cambrensis 99*, 57-63

1948 The copper mines on the Great Ormes Head, Caernarvonshire, *Archaeologia Cambrensis 100*, 61-66

Davies, W. 1815 The Agriculture of South Wales: Vols.1& 2

Dawkins,W.B. 1875 On the stone mining tools from Alderley Edge Cheshire, *Proc.Lit. & Phil.Soc. Manchester, 14*, p.74-79. Reprinted in *Journ. of the Anthropological Institute of Great Britain, 5* (1876)

Dutton, L.A. & Fasham, P. 1994 Prehistoric copper mining on the Great Orme, Llandudno, Gwynedd, *Proceedings of the Prehistoric Soc.*, Vol.60, 245-286

Earl, B. 1968 *Cornish Mining (The techniques of metal mining in the West of England, past and present)*, Bradford Barton, Truro, 119 pp

Ellis-Jones, J. 1988 *Aspects of Ancient Metallurgy: Acta of a British School at Athens Centenary Conference at Bangor, 1986*. Bangor, University College of Wales, 154 pp

English Heritage 1998 *Dendrochronology: guidelines on producing and interpreting dendrochronological dates*, London

Evans, Sir John 1897 *The Ancient Stone Implements, Weapons and Ornaments of Great Britain*, 2nd.Edition, London, 233-236

Flanagan, L.N.W. 1980 Industrial sources, production and distribution in Early Bronze Age Ireland, *The Beginnings of Metallurgy in Atlantic Europe*, ed. Ryan, M., Dublin, 145-163

Ford, T.D. 1998 Primitive Mining Tools from Tembelini nr. Syama, Mali, W.Africa, *Mining History - The Bull. Peak Dist.Mines Hist.Soc. 13 (6)*, 6

Ford, T.D. & Willies, L. 1994 Mining Before Powder, *Bull. Peak District Mines Hist. Soc., Vol.12, no.3*, 149 pp

Francis, A. 1874 *History of the Cardiganshire Mines from the earliest ages, and the authenticated history to A.D. 1874, with their present position and prospect*, publ. Goginan, Aberystwyth. Reprinted by Mining Facsimiles, No.14, Sheffield, 1987, 147 pp.

Frost, P. & Hankinson, R. 1995 Lllangollen, Craig y Forwyn, World's End, *Archaeology in Wales 35*, 45

Gale, D. 1986 *Recording an elevation of a copper mining face at EngineVein, Alderley Edge*, Undergraduate Thesis, School of Archaeological Sciences, University of Bradford, 322pp

1989 Evidence of ancient copper mining at Engine Vein, Alderley Edge, *Bull.Peak Dist. Mines Hist.Soc., Vol.10(5)*, 266-273

1995 *Stone tools employed in prehistoric mining*, unpubl. PhD Thesis, Department of Archaeological Sciences, University of Bradfod, 322 pp.

Gale, D. & Ottaway, B.1990 An early mining site in the Mitterberg ore region of Austria, in Crew & Crew (eds.), 36-38

Gale, N.H. & Stos-Gale, Z.A. 1981 Cycladic lead and silver metallurgy, *Annals of the British School at Athens 76*, 169-224

Glumac, P.D. & Todd, J.A. 1987 New evidence for the use of lead in prehistoric SE Europe, *Archaeomaterials 2*, 29-37

Goldenberg, G. & Reiser, B. 1995 Late Bronze Age fahlore-mining and extractive metallurgy in the area of Shwarz/Brixlegg in N.Tyrol, Austria, Abstracts: An International Workshop of the Univ. Innsbruck '*Urgeschichtliche Kupfergewinnung in Alpenraum*' (unpubl.)

Goodburn, D. *(forthcoming)* The woodwork evidence from Swalecliffe, Kent, *Proceedings of Prehistoric Society*

Goodburn, D. & Minkin, J.1997 The Woodworking Evidence from MOLAS site - Bull Wharf, *Assessment Report (1997)*, Museum of London (unpubl.)

Griffin, J.B. 1961 Early American Mining in Keewanaw Peninsula, Lake Superior, *Anthropological Papers No.17*, Museum of Anthropology, University of Michigan Press, Michigan U.S.A.

Guilbert, G. 1996 The oldest artefact of lead in the Peak: new evidence from Mam Tor, *Mining History : Bull. Peak Dist. Mines Hist.Soc. 13 (1)*, 12-18

Harding, A.F. 1985 *Climate Change in Later Prehistory*, Edinburgh

Hauptmann, A. 1990 The copper ore deposit of Feinan, Wadi Araba: early mining and metallurgy, IN *The Near East in Antiquity 1*, Kerner, S. (ed.), Goethe-Institut, Amman, 53-62

HindzincTech 1989 Zinc - A Heritage, *Hindzinc Tech 1 (1)*, Jan. 1989, 12pp

Hedges, R.E.M., Law, I.A., Bronk, C.R., and Housley, R.A. 1989 The Oxford accelerator mass spectrometry facility: technical developments in routine dating, samples combustion to carbon dioxide, and in the Oxford AMS carbon dioxide ion source system, *Archaeometry 31(2)*, 99-114

Hedges, R.E.M., Housley, R.A., Law, I.A. & Bronk, C.R. 1990 Radiocarbon dates from the Oxford AMS system: datelist 10, *Archaeometry 32 (1)*, 101-108

Henshall, A.S. 1979 Artefacts from the Quaterness Cairn, IN Renfriw, C. (ed.) *Investigations in Orkney*, Soc.Antiquaries, London, 75-93

Herbert, E.W. 1984 *Red Gold of Africa - Copper in Precolonial History and Culture*, The University of Wisconsin Press, 413 pp.

Holgate, R. 1991 *Prehistoric Flint Mines*, Shire Archaeology no.67, 56 pp.

Holzer, H.F. & Momenzadeh, M. 1971 Ancient copper mines in the Veshnoveh area, Kuhestan-e-Qom, West-Central Iran, *Archaeologia Austriaca 49*, 1-22

Houlder, C.H. 1957 The excavation of a barrow in Cardiganshire, *Ceredigion 3*, 11-23 & 118-123

Hughes, S.J.S. 1981 *The Cwmystwyth Mines*. British Mining No.17, Northern Mine Research Soc. Monograph, Sheffield, 78 pp

Humphrey, S. 1989 Petrology and provenance of a pebble hammer, Cwmystwyth, *Archaeology in Wales 29*, 43

Hunt, R. 1848 Notices of the history of the Lead Mines of Cardiganshire, *Memoirs of the Geol. Surv. G.B., Vol.2, Part 2*, 535-653

Ixer, R.A. 1997 *Report on the Copa Hill mineralization*. In Geoarchaeological research into Prehistoric mining for copper in Wales. D.A. Jenkins and S. Timberlake. Unpublished report to the Leverhulme Trust. Appendix 2, 113-114

Ixer, R.A. and Budd. P. 1998 *The mineralogy of Bronze Age copper ores from the British Isles: implications for thr composition of early metalwork*. Oxford Journal of Archaeology, 17, 15-41

Jackson, J.S. 1968 Bronze Age copper mining on Mt.Gabriel, West Co.Cork, Ireland, *Archaeologia Austriaca, Vol.43*, 92-114

1980 Bronze Age copper mining in counties Cork and Kerry, Ireland, IN Craddock, P. (ed.) 1980, *Scientific Studies in Early Mining and Extractive Metallurgy*, 9-30

James, D. 1988 Prehistoric copper mining on the Great Orme's Head IN Ellis-Jones, J. (ed.), 1988, 115-121. Reprinted in Crew & Crew (eds.), 1990, 1-4

James, D.B. 2001 *Ceredigion, Its Natural History*, publ. James D.B., Bowstreet, Ceredigion, Cambrian Printers, Aberystwyth, 204 pp

Jenkins, D.A. 1995 Mynydd Parys Copper Mines, *Archaeology in Wales 35*, 35-37

Jenkins, D. & Timberlake, S. 1997 *Geoarchaeological research into prehistoric mining for copper in Wales*, A Report to the Leverhulme Trust, University of Wales, Bangor (unpubl.), 141 pp

Jones, G.D.B. 1979 The Roman Evidence, IN Crawford, H. (ed.) 1979, 85-99

Jones, O.T., 1922 Lead and zinc. The Mining District of North Cardiganshire and West Montgomeryshire, *Memoirs of the Geological Survey: Special Reports on the Mineral Resources of Great Britain 20*, London, HMSO, 205 pp

Jones, S.G. 1997a Pentrwyn Metal Working Site, Great Orme, Project G1497, *Gwynedd Archaeological Trust Report no. 321*, March 1999, unpubl.

1997b Great Orme Bronze Age Smelting Site, Llandudno, *Archaeology in Wales 39*, 79

Jones, S.L. 2001 *The environmental impact of Bronze Age mining on Copa Hill, Cwmystwyth, mid-Wales*. Unpublished BSc (hons) thesis, Coventry University

Jovanovic, B. 1979 The technology of primitive copper mining in south-east Europe, *Proc. Prehistoric Soc. 45*, 103-110

Knapp, A.B. 1999 The Archaeology of Mining: Fieldwork Perspectives from the Sydney Cyprus Survey Project (SCSP), IN *Metals in Antiquity*, eds. Young, S.M.M., Pollard, A.M., Budd, P. & Ixer, R.A., *BAR International Series 792*, 98-109, Oxford

Krumbein, W.C. 1941 Measurement and geological significance of shape and roundness of sedimentary particles, *Journal of Sedimentary Petrology 11 (2)*, 64-72

Kyrle, G. & Klose, O. 1918 Urgeschicte der Osterrich Kronlander Salzburg, Osterrich, *Kunstopographie XVII*, Wein 1918

Lechelon, B. 1974 La Mine Antique de Bouche-Payrol (Sud Aveyron), *Essai d'archeologie miniere de la Narbonaise*, ed. Fayet, 61 pp.

Leighton, D.K. 1984 Structured round cairns in west central Wales, *Proc.Prehist.Soc. 50*, 319-350

Leland, John 1536-1539 Itinerary in Wales. IN (part) Meyrick, S.M. 1907 edition

Lewis, A. 1990a Underground exploration of the Great Orme Copper Mines IN Crew & Crew (eds) 1990, 5-10

1990b Firesetting experiments on the Great Orme's Head, 1989, IN Crew, P. & Crew, S. (eds.) *Early Mining in the British Isles* (1990), 55-56

1994 Bronze Age mining on the Great Orme IN Ford & Willies (eds.), 1994, 31-36

1996 *Prehistoric mining at the Great Orme: criteria for the identification of early mining*, unpublished MPhil. thesis, University of Wales, Bangor, 183 pp

Lewis, P.R.1977 The Ogofau Roman Gold Mines at Dolaucothi, reprinted from *The National Trust Year Book 1976-77*, publ. for The National Trust by Europa Publications., 20-35

Liscombe & Co. 1869/1870 *The Mines of Cardiganshire, Montgomeryshire and Shropshire*, publ. Liverpool. Reprinted Hughes, S.J.S., Talybont, 1989, 45 pp

Lucht, W.H. 1987 Die Käfer Mitteleuropas, Katalog Krefeld: Goecke & Evers.

Lynch, F. 1991 *Prehistoric Anglesey* (2nd.edition), Anglesey Antiquarian Society, Llangefni, 411 pp

Maddin, R. (ed.) 1986 *The Beginning of the Use of Metals and Alloys*, proceedings of 2nd conference on metals and alloys, Zengzhou, China, October 1986, publ. MIT, Cambridhe Mass.

Maggi, R. & Del Lucchese, A. 1988 Aspects of the Copper Age in Liguria, L'eta del rame nell'Italia Centrale, *Rassegna di Archeologia estratto 7*, 331-339

Marshall, P.D., O'Hara, S. L. & Ottaway, B.S. 1999 Early copper metallurgy in Austria and methods of assessing its impact on the environment, IN The Beginnings of Metallurgy, *Der Anschnitt 9*, eds. Hauptmann, A., Pernicka, E., Rehren, T. & Yalcin, U., Bochum, 255-266

Mason, J.S. 1994 *A regional paragenesis for the Central Wales Orefield*. Unpublished M.Sc. thesis, University of Wales

1996 Precious and strategic metals in an old Pb-Zn mining district: the Central Wales Orefield re-appraised IN Mineralisation in the Caledonides (Edinburgh: Institute of Mining and Metallurgy, Edinburgh Geol. Soc., Irish Association of Economic Geology)

1997 *Regional polyphase and polymetallic vein mineralisation in the Caledonides of the Central Wales Orefield.* Transactions of the Institution of Mining and Metallurgy. (Section B Applied Earth Science) 106, B135-144

Meyrick, S.M. 1808 *The History and Antiquities of the County of Cardigan* (1907 edition), National Library of Wales ms.

Mighall, T.M. & Chambers, F.W., 1993a The environmental impact of prehistoric mining at Copa Hill, Cwmystwyth, Wales. *The Holocene* 3, 260-4

1993b Early mining and metalworking: its impact on the environment, *Journal of the Historical Metallurgy Society 27 (2)*, 71-83

1994 Vegetation history and Bronze Age mining on Mt.Gabriel: preliminary results, IN *Mount Gabriel - Bronze Age mining in Ireland*, O'Brien, W.F. (1994), Appendix 4: 289-298

Mighall, T.M, Grattan, J.P., Forsyth, S. & Timberlake, S. 2000a Tracing atmospheric metal mining pollution in blanket peat' IN *Tracers in Geomorphology* Foster,D.L.(ed.), publ. J.Wiley, 101-121

2000b Bronze Age lead mining at Copa Hill, Cwmystwyth - fact or fantasy?, *Historical Metallurgy 34(1)*, 1-12

Mighall, T.M., Abrahams, P.W., Grattan, J.P., Hayes, D., Timberlake, S. & Forsyth, S. 2002a (in press) Geochemical evidence for atmospheric pollution derived from prehistoric copper mining at Copa Hill, Cwmystwyth, mid-Wales, U.K., IN *Science of the Total Environment*

Mighall, T.M., Grattan, J.P., Lees, J.A., Timberlake, S. & Forsyth, S. 2002b (in press) An atmospheric pollution history for lead-zinc mining from the Ystwyth Valley, Dyfed, mid-Wales, UK as recorded by an upland blanket peat, IN *Geochemistry*

Mighall, T.M., Timberlake, S., Clark, S.H.E. & Caseldine, A. 2002c (in press) A palaeoenvironmental investigation of sediments from the prehistoric mine of Copa Hill, Cwmystwyth, mid-Wales, IN *Journal of Archaeological Science*

Moore, P.D. 1968 Human influence upon vegetation history in N.Cardiganshire, *Nature 217*, 1006-1009

Morris, L. 1751 *A short history of the Crown Manor of Creuthyn in the County of Cardigan, South Wales* IN Meyrick, S.M. 1808

1756 *History of the Manor of Mevenith*, Chapter 13-14 in Ms copy, National Library of Wales

Murphy, K.1987 Plas Gogerddan, *Archaeology in Wales 26*, 29-31

Mussche, H.F., Bingen, J., Jones, J.E., Waelkens, M. 1990 *Thorikos IX 1977/1982 Preliminary Report*, publ. Comite des Fouilles Belges en Grece, Gent, 143 pp.

Needham, S.P., Lawson, A.J. & Green, H.S. 1985 *Early Bronze Age Hoards*, British Bronze Age Metalwork A1-A6, London

Newman, P. (ed.) 1996 'The Archaeology of Mining and Metallurgy in South West-Britain', *Mining History: Bulletin Peak District Mines Historical Soc. 13 (2)*, 167 pp

Northover, J.P. 1980a Bronze in the British Bronze Age IN Oddy, W.A. 1980, 63-70

1980b The analysis of Welsh Bronze Age metalwork: Appendix in *Guide Catalogue of the Bronze Age Collections*, Savory, H.N., National Museum of Wales, 229-243

1982 The exploration of the long-distance movement of bronze in in Bronze and Early Iron Age Europe, *Institute of Archaeology Bulletin 19*, 45-72

1983 The exploration of the long distance movement of bronze in Bronze Age and Early Iron Age Europe, *Bull. Institute of Archaeology (1983)*, University of London, 45-72

1999 The earliest metalworking in Southern Britain IN 'The Beginnings of Metallurgy' (eds. Hauptmann, A., Pernicka, E., Rehren, T., Yalcin, U.) *Der Anschnitt 9* Bochum, 211-226

2002 (*forthcoming*) History of Bronze Analysis in Britain and the use of data today IN *The Beginnings of Metallurgy in the Old World*, Proceedings of Euroseminar Conference in Freiberg, Germany, November 1999, Bartelheim, M., Pernicka, E. & Krause, R. (2002) (eds.)

O'Brien, W. 1987 The dating of the Mt.Gabriel - type copper mines of West Cork, *Journal of the Cork Historical and Archaeological Soc. Vol.92*, 50-70

1989 A primitive mining complex at Derrycarhoon, County Cork, *Journal of the Cork Historical and Archaeological Society 94 (253)*, 1-18

1990 Prehistoric copper mining in south-west Ireland, *Proceedings of the Prehistoric Society 56*, 269-290

1994 *Mount Gabriel: Bronze Age Mining in Ireland*, Galway, Galway University Press, 371 pp

1995 Ross Island and the origins of Irish-British metallurgy IN *Ireland in the Bronze Age*, Proceedings of the Dublin Conference April 1995 (eds. Waddell, J. & Shee Twohig, E.), 38-48

1996 *Bronze Age Copper Mining in Britain and Ireland*, Shire Archaeology No.71, 64 pp

1998 La mine de cuivre de Ross Island et la metallurgie chalcolithique en Irlande, *Paleometallurgie des cuivres*, Actes du colloque de Bourg-en-Brasse et Beaune, France, October 1997, 101-107

1999 Resource availability and metal supply in the insular Bronze Age, The Beginnings of Metallurgy, *Der Anschitt 9*, (eds. Hauptmann, A., Pernicka, E., Rehren, T. & Yalcin, U.), 227-236, Bochum

O'Sullivan, A. 1997 Neolithic, Bronze Age and Iron Age Woodworking Techniques IN Rafferty, B., Trackway Excavations in the Mount Dillon Bogs, Co.Longford, Eire, *Irish Archaeological Wetland Transactions 3*, 291-342, Dublin

Oddy, W.A.(ed.) 1980 Aspects of Early Metallurgy, *British Museum Occ. Paper 17*, London

Ottaway, B. & Wager, E. 2000 Ffynnon Rufeiniog, Great Orme, Llandudno, *Archaeology in Wales 40*, 73

Owen, T.M. 1969 Historical aspects of Peat-Cutting in Wales IN Jenkins, G. (ed.) *Studies in Folk Life: Essays in honour of Iorwerth C Peate*, London: RKP, 123-156

1975 Historical aspects of Peat-Cutting in Merioneth, *Journal of the Merioneth Historical and Record Society 7*, 309-321

Pain, S. 2001 'Legends of the Edge', *New Scientist*, 4th.August 2001, 44-45

Peake, H.J.E. 1937 Chairman's Report of the Committee appointed to investigate Ancient Mining in Wales IN *Report of the British Association for the Advancement of Science 1937*, 301-303

Pearce, S.M. 1984 Bronze Age Metalwork in Southern Britain, *Shire Archaeology No.39*, Aylesbury, 64 pp

Pearson, G.W. & Stuiver, M. 1986 High precision calibration of the radiocarbon time scale, 500 - 2500 BC, *Radiocarbon 28*, 839-86

Penhallurick, R.D. 1986 *Tin in Antiquity*, The Institute of Metals, London, 271 pp.

Phillips, W.J. 1972 Hydraulic fracturing and mineralisation, *Journal of Geol. Soc. London 128*, 337-359

Pickin, J. 1988 Stone tools and early mining in Wales, *Archaeology in Wales 28*, 18-19

1990 Stone tools and early metal mining in England and Wales IN Crew & Crew (eds.) 1990, 39-42

1999 Stone hammers from the Ecton Mines in the Bateman Collection, Sheffield, *Mining History: Bull. Peak Dist. Mines Hist. Soc., Vol.13, no.5*, 15-18

Pickin, J. & Timberlake, S. 1988 Stone hammers and fire-setting: a preliminary experiment at Cwmystwyth Mine, Dyfed, *Bull. Peak Dist. Mines Hist. Soc. 10 (5)*, 274-275

Pittioni, R. 1951 Prehistoric copper mining in Austria: problems and facts, *Institute of Archaeology Seventh Annual Report*, 16-43

Prag, J., Garner, A. & Housley, R. 1991 The Alderley Edge shovel: an epic in Three Acts, *Current Archaeology 137*, 172-175

Rafferty, J. 1961 National Museum of Ireland archaeological acquisitions in 1959, *Journal Royal Soc. Antiq. Ireland 61*, 43-10

Rieser, B. & Schrattenthaler, H. 1995 Prehistoric Mining tools in the area of Shwarz/Brixlegg in North Tyrol, Austria, Abstracts of an International Workshop of the University of Innsbruck *Urgeschichtliche Kupfergewinnung in Alpenraum*, October 1995

Richardson, J.B. 1974 *Metal Mining*, Industrial Archaeology Series (general ed. L.T.C.Rolt), Allen Lane, London, 207 pp

Roeder, C. & Graves, F.S. 1905 Recent archaeological discoveries at Alderley Edge, *Trans. Lancs. Cheshire Antiquarian Soc. 23*, 17-21

Rohl, B. & Needham, S. 1998 The circulation of metal in the British Bronze Age: the application of lead isotope analysis, *British Museum Occasional Paper no.102*, 234 pp

Rostoker, W., Pigott, V.C. & Dvorak, J.R. 1989 Direct eduction to copper metal by oxide-sulfide mineral interaction, *Archaeomaterials 3(1)*, 69-87

Rothenberg,B. & Blanco-Freijeiro, A. 1980 Ancient copper mining and smelting at Chinflon (Huelva, SW Spain). IN Craddock, P. (ed.), *Scientific Studies in Early Mining and Extractive Metallurgy*

1981 *Studies in Ancient Mining and Metallurgy in South-West Spain*, London, Institute for Archaeo-Metallurgical Studies (UCL), 320 pp

Savory, H.N. 1980 *Guide Catalogue of the Bronze Age Collections*, National Museum of Wales, Cardiff, 258 pp

Schweingruber, F. 1978 *Microscopic Wood Anatomy*

Sharpe, A. 1997 Smoke but no fire, or the mysterious case of the missing miner, IN Budd & Gale (eds.) 1997, 35-40

Shell, C.M. 1979 The early exploitation of tin deposits in SW England, IN Ryan, M. (ed.) *The Origins of Metallurgy in Atlantic Europe*, Dublin: Stationary Office, 251-263

Shennan, S. 1999 Cost, benefit and value in the organization of early European copper production, *Antiquity 73 no.280* (June 1999), 352-363

Shepherd, R. 1980 *Prehistoric Mining and Allied Industries*, Academic Press, London

Sieveking, G. de G.1979 Grime's Graves and prehistoric European flint mining IN Crawford, H. (ed.) *Subterranean Britain*, 1-43

Smyth, W.W. 1848 On the Mining District of Cardiganshire and Montgomeryshire, *Memoirs Geological Survey, Vol.2, Part 2*, 655-684

Spargo, T. 1870 *The Mines of Wales, their present position and prospects*, publ. London. Reprinted by Hughes, S.J.S., Y Lolfa, Talybont, 1975, 77 pp

Stace, C. 1991 *New Flora of the British Isles*. Cambridge University Press

Stuiver, M. & Polach 1977 *Radiocarbon 19*, 355-363

Stuiver, M., Reimer, P. J., Bard, E., Beck, J. W., Burr, G. S., Hughen, K. A., Kromer, B., McCormac, G. van der Plicht, J. and Spurk, M. 1998 INTCAL98 radiocarbon age calibration, *Radiocarbon, 40(3)*, 1041-1084

Switsur, R. 1981 Cambridge University Natural Radiocarbon Measurements XV, *Radiocarbon 23 (1)*, 81-93.

Sykes, C. Sir 1796 *Journal of a Tour in Wales*, National Library of Wales, Ms 2258C

Tabor, R. 1994 *Traditional Woodland Crafts*, Batsford, London, 111 pp

Taylor, J.A. 1973 Chronometers and chronicles: a study of palaeoenvironments in west-central Wales, *Progress in Geography 5*, 247-334

Taylor, M. 1992 Flag Fen: The Wood, *Antiquity 66*, 476-498

Timberlake, S., 1987 An archaeological investigation of early mineworkings on Copa Hill, Cwmystwyth, *Archaeology in Wales 27*, 18-20

1988 Excavations at Parys Mountain and Nantyreira, *Archaeology in Wales* 28, 11-17

1990a Excavations at an early mining site on Copa Hill, Cwmystwyth, Dyfed, 1989 & 1990, *Archaeology in Wales* 30, 7-13

1990b Excavations and fieldwork on Copa Hill, Cwmystwyth, Dyfed, 1989, in Crew & Crew (eds.), 1990, 22-29

1990c Firesetting and primitive mining experiments, Cwmystwyth, 1989, IN Crew, P. & Crew, S. (eds.) *Early Mining in the British Isles*, 53-54

1990d Review of the historical evidence for the use of firesetting, IN Crew, P. & Crew, S. (eds.), 49-52

1991 New evidence for early prehistoric mining in Wales - problems and potentials IN *Archaeological Sciences 1989*, Proceedings of Bradford Conference September 1989 (eds. Budd, P., Chapman, B., Jackson, C., Janaway, R. & Ottaway, B.), 179-193

1992a Prehistoric copper mining in Britain, *Cornish Archaeology 31*, 15-34

1992b Llancynfelin Mine, *Archaeology in Wales 32*, 90-91

1993 Cwmystwyth: 3500 years of mining history - some problems of conservation and recording, *Journal of the Russell Society 5 (1)*, 49-53

1994a Archaeological and circumstantial evidence for early mining in Wales IN Ford & Willies (eds.) 1994, 133-143

1994c An experimental tin smelt at Flag Fen, *The Journal of Historical Metallurgy Society Vol.28, no.2*, 122-129

1995a Copa Hill, Cwmystwyth, *Archaeology in Wales 35*, 40-43

1995b Llancynfelin and Nantyrarian Mines, *Archaeology in Wales 37*, 62-65

1996a Copa Hill, Cwmystwyth, *Archaeology in Wales 36*, 60-61

1996b Tyn y fron Mine, Cwmrheidol, *Arch. in Wales 36*, 61-63

1998 Survey of early metal mines within the Welsh Uplands, *Archaeology in Wales 38*, 79-81

2002a Mining and prospection for metals in Early Bronze Age Britain - making claims within the archaeological landscape, IN Bruck, J. (ed.) *'Bronze Age Landscapes - Tradition and Transformation'*, publ. Oxbow Books, 179-192

2002b Ore prospection during the Early Bronze Age in Britain, IN Bartelheim, M., Pernicka, E. & Krause, R. (eds.) 'The Beginnings of Metallurgy in the Old World', *Archaometrie – Freiberger Forschungen zur Altertumswissenschaft 1*

2003 (forthcoming) Archaeological Excavations at Engine Vein, Alderley Edge IN The Archaeology of Alderley Edge, *BAR Series*, BAR Publishing, Oxford

Timberlake, S. & Mason, J. 1997 Ogof Wyddon (Machynlleth Park Copper Mine), *Archaeology in Wales 37*, 62-65

Timberlake, S. & Mighall, T.M. 1992 Historic and prehistoric mining on Copa Hill, Cwmystwyth, *Archaeology in Wales 32*, 38-44

Timberlake, S. & Switsur, R. 1988 An archaeological investigation of early mineworkings on Copa Hill, Cwmystwyth: New Evidence for Prehistoric Mining, *Proceedings of the Prehistoric Society 54*, 329-333

Todd, M. 1996 Ancient Mining on Mendip, Somerset: A Preliminary Report on Recent Work IN Newman, P. (ed.), *Mining History 13*, 47-51

Tyers, I. 1999 Dendro for Windows program guide 2nd edition, *ARCUS Rep. 500*

Tylecote, R.F. 1962 *Metallurgy in Archaeology*, Arnold, 368 pp

1986 *The Prehistory of Metallurgy in the British Isles*, The Institute of Metals, London, 257 pp

Tyler, A.W. 1982 *Prehistoric and Roman Mining for Metals in England and Wales*, PhD thesis, University of Wales, Cardiff, 314 pp

Vermeersch, P. & Paulissen, E. 1989 The oldest quarries known: Stone Age miners in Egypt, *Episodes 12 (1)*, 1989, 35-36

Wagner, G.A., Gentner, N., Gropengiesser, H. & Gale, N.H. 1980 Early Bronze Age lead-silver mining and metallurgy in the Aegean: The Ancient workings on Siphnos IN Craddock, P. (ed.) 1980, 63-86

Waller, W. 1699 *A description of the mines of Cardiganshire* (IN Bick 1991,64; Hughes 1981,12)

Walters, M. 1994 *Powys metal mines survey 1993 - Report No.89*, Clwyd-Powys Archaeological Trust, Welshpool, unpubl.

Warrington, G. 1981 The Copper Mines of Alderley Edge and Mottram St.Andrew, Cheshire, *Journal of the Chester Archaeological Society 64*, 47-73

Watson, E. 1968 The periglacial landscape of the Aberystwyth region IN Bowen, E.G., Carter, H. & Taylor, J.A. (eds.), *Geography at Aberystwyth: Essays written on the occasion of the Departmental Jubilee*. University of Wales Press, Cardiff, 35-49

White, R.B. 1977 Rhosgoch to Stanlow Shell Oil Pipeline, *Bull. Board Celtic Studies 27* (1976-1977), 463-496

Williams, J. 1780 *The Natural History of the Mineral Kingdom*, Volume 1

Williams, J.G. 1866 *A short account of the British Encampments lying between the Rivers Rheidol and Llyfnant in the County of Cardigan and their connection with the mines*, Aberystwyth, Cambrian Press, 56 pp

Williams, K.E. 1927 The Glacial Drifts of Western Cardiganshire, *Geological Magazine 44*, 205-227

Williams, S.W. 1889 *The Cistercian Abbey of Strata Florida*

Willies, L. 1990 An Early Bronze Age Tin Mine in Anatolia, Turkey, *Bull. Peak Dist. Mines Hist. Soc. 11 (2)*, 91-96

Worthington, T. & Craddock, B.R. 1996 Modern Stone Tools, *Mining History - Bull. Peak Dist. Mines Hist. Soc. 13 (1)*, 58

APPENDIX 1

EMRG STANDARD RECORD SHEET FOR HAMMER-STONES (after Jenkins & Timberlake 1997)

1: IDENTITY

Sample No.	Year	Sector	Context (Δ)
CH 99:102 h	1999	D7 Deep Fssr 1	100

2: DIMENSIONS etc.

Weight (g)	Length x (cm)	Width y (cm)	Depth z (cm)
1500	23	13	10.5

3: SHAPE etc.

Original shape	Natural indentation	Roundness factor (1-9)	Surface (1-4) smoothness	Natural/modern abrasion	Hardness (3-8 mohs)
5 Rectangular ✓	✓		Smooth (shore)	Glacial striae	3
1 Spherical		6	1	Beach attrition ✓	4
3 Ovoid ✓		(7)	(2)	Other	5
2 Pear		8	3		6
4 Cylindrical		9	4	Modern	(7)
6 Flattened			Rough	(accidental)	8

4: PETROLOGY

	Grain size	Sorting	Banding	Grain roundness	Fels. %	Matrix cement	Structure
Siltstone				0 (angular)			
(Greywacke)	Fine	v. well		1	<2		Joints
Sandstone	(Medium)	Well		2	2-5		Bedding
Arkose	Coarse	(Moderate)		(3)	(5-10)		Veins
Quartzite	v. coarse	Poor		4	10-25		Cleavage
Other	Pebbly			5	25-33		(Other)
				6 (spherical)	33-50		quartz vein

5: COMPLETENESS

Small	<25%	Unused
Spall	25-40%	Pebble
	40-60%	
	60-75%	
	(>75%)	

6: TYPE OF FRACTURE

(Conchoidal)	Good
	(Medium)
	Poor
Uneven	
Planar	(Premature fracture)

7: WEAR ANALYSIS – 1 Hammering

	End wear			Edge wear		Side wear	
Single ended (1)		Double ended (2)					
Primary	secondary	Primary	secondary	Primary	secondary	primary	secondary
Bruising				Bruising		Slight bruising	
Slight pounding			✓			Bruising ✓	
Moderate pounding		✓			✓	Indentation	
Heavy pounding				Heavy pounding	Heavy pounding		
Flaking		✓					
Heavy flaking							
Faceted area				Faceted area		Faceted area	
Mineral residue		Mineral residue		Mineral residue		Mineral residue	

7.2 Grinding		7.3 Crushing anvil use		7.4 Re-use as flake
Location	Primary	Secondary		
Faint	Slight bruising			Hammering end
	Strong bruising			Chisel end
Strong	Indentation			
				Degree of use
Grinding marks – striations	Location	✓		Slight
	Flake surface			Moderate
Faceted area/	Pebble surface			Heavy
Mineral residue	Mineral residue	✓		Scratch markings

8: MODIFICATION (for hafting/handgrip)

Position	Primary				Secondary			Contemporary scratch/ Scoring	Evidence for haft wear?
	Edges	Side	Semi-continuous	groove	Edges	side	Use of Natural indentation in Pebble		
mid girth ✓	✓								
Towards narrow end	Scratch notching				Rounded sharp ✓ edge of Flake			certain	
Towards broad end									
Centre of gravity					Notching			Possible	

9: SUMMARY DESCRIPTION OF TOOL
(Estimate)
- (well used)
- poorly used
- long survival of usefulness
- (hammering)
- (crushing)
- small hammer for other tool
- (anvil stone)
- re-used
- recommend for detailed drawing ✓
- recommend retain collection ✓

10: PHOTOGRAPHS
B+W (x3 views)

11: DRAWINGS
(outline profiles -annotated)
longitudinal / t.s.

HAMMER-STONE DATA RECORD KEY:

COBBLE SHAPE: spherical=1; pear=2; ovoid=3; cylindrical=4; rectangular=5; flattened=6

COBBLE ROUNDNESS (1 to 9): factor 6 (slight irreg) - 9 (v round)

COBBLE SMOOTHNESS (1 to 6): 1 (v smooth = beach pebble) - 6 (rough)

SURFACE ABRASION: GS=glacial striae; BA=beach attrition; M=modern (accidental etc)

HARDNESS: Moh's hardness scale   2 (soft) - 8 (v hard)   NB vein quartz = 7-7.5

PETROLOGY (rock type): A= mudstone/siltstone; B= laminated fine grained flaggy sandstone; C =fine grained felspathic grit/greywacke; Ca = more felspathic type; Cc=micaceous greywacke/bedded flagstone; Cb= quartzitic greywacke; D = quartzitic grit; E= quartzite; F= exotic(glacial erratic cobble - incl non-local quatzites and igneous rocks eg. dolerites, qtz porphyry)
Identified types of all cobbles include: Types C, Cb, Cc, Ca, D, Cc/Cb, Cb/D, Ca/Cc,Ca/Cb, Ca/D, Cc/D, A, B, E, F

GRAIN SIZE (1-5): 1(Fine grained) - 5 (small pebbly clasts)

GRAIN ROUNDNESS FACTOR (1to 6): 1(angular) - 5 (nr spherical/well rounded grains)

FELDSPAR CONTENT: % of felspar from <5% - >75% (white grains of detrital plagioclase or orthoclase)

STRUCTURE: J=joints; B=bedding; QV = quartz vein; CL=cleavage

COMPLETENESS = % of cobble surviving, either as core or flake/spall

TYPE OF FRACTURE: C=conchoidal; U=uneven; PL=planar (usually premature fracture along joint plane or qtz vein

FRACTURE SCALE: 1=premature(planar)fracture; 2=uneven; 3=poor conchoidal; 4=medium conchoidal; 5= good conchoidal

WEAR ANALYSIS: END WEAR: x2=double ended; primary + 2ndry = use/re-use; 1=bruising; 2=slight pounding; 3=moderate pounding; 4=heavy pounding; 5=flaking; 6=heavy flaking (similar scale with EDGE WEAR & SIDE WEAR. FACET: x1 or x2 etc facets or surfaces worn flat from wear

CRUSHING ANVIL USE: flk= location on flake or fracture surface; pbl = location of use on natural pebble surface; 1=slight bruising; 2= strong bruising; 3=indentation

RE-USE AS FLAKE TOOL: H=hammering use; CH=chisel/pick use; 1=slight use; 2=moderate use; 3=heavy

MODIFICATION OF COBBLE (HAFTING): POSITION: A=mid-girth; B=towards narrow end; C=towards broad end; D=centre of gravity;
TYPE OF MODIFICATION: 1=scratch notching; 2=edge notches in cobble; 3=side notch/indentation; 4=semi-continuous groove; A= secondary notching of edges; B=rounded off sharp edges of flake; C= 2ndry notching; D=2ndry side notch

HAMMER TYPE CLASSIFICATION: **1A**= unmodified single end use as hammer; **1Aa**=double end use (1 end >); **1B**=use of side pebble as hammer for crushing/or as mallet; **1C**=broken tip of hammer (or small pebble) re-used as hand-held crushing/pounding tool around edges; **1D**= flake tool re-use (as chisel etc); **1E**= use/re-use of cobble surface as anvil stone (for crushing); **1F**= pecking stone for notching (typically lump quartz)

2A = modified (notched for hafting) single end hammer; **2Aa** = ...double end hammer;
2AE = modified end hammer/anvil stone;   **3A** = further modified hammer(semi-grooved for hafting) ...etc

KRUMBEIN's (1941) visual roundness chart for cobbles:

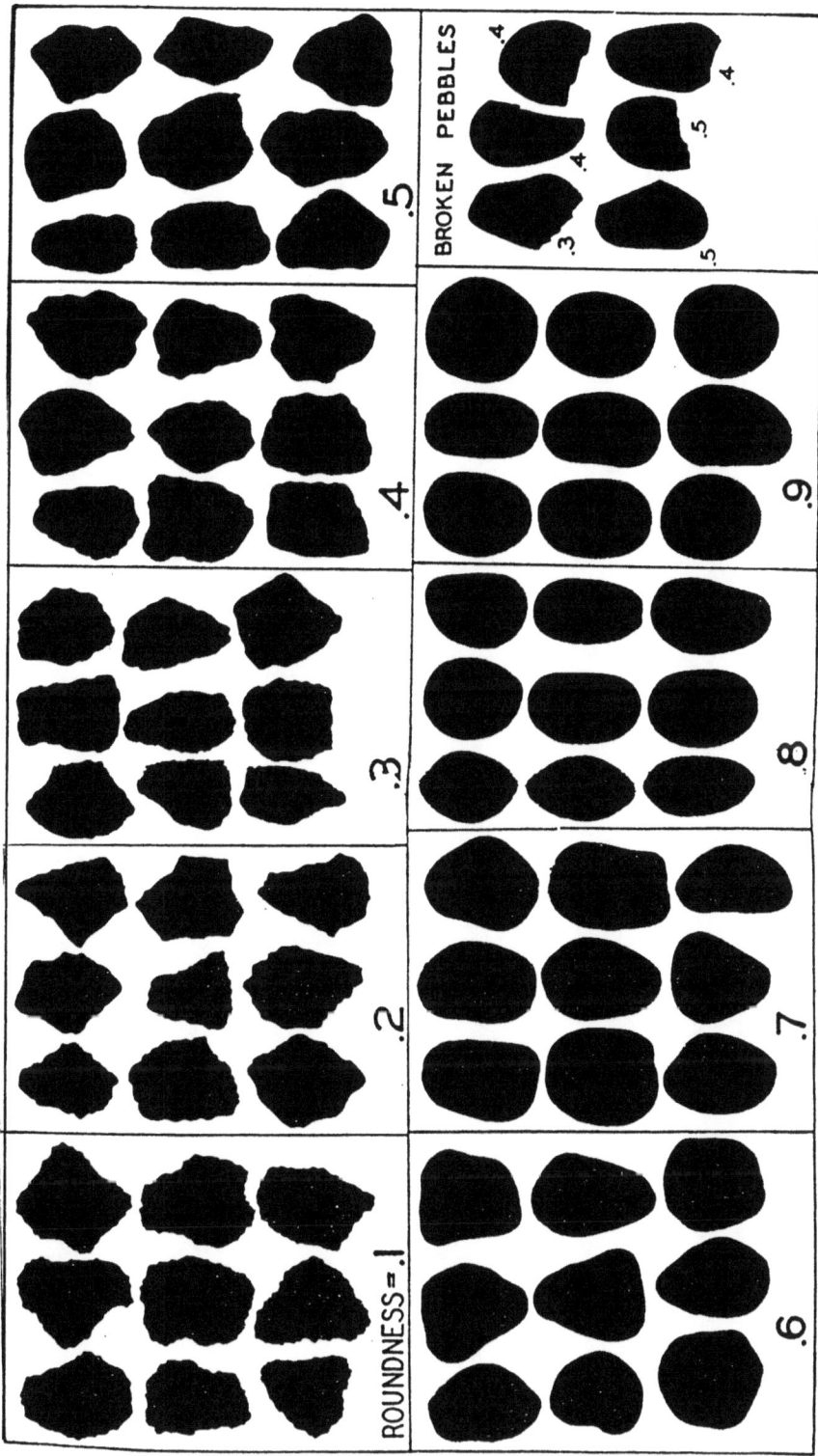

Krumbein's (1941) visual roundness chart for classifying original cobble shape (with EMRG Standard Record Sheet)